Research Report

Extreme Cost Growth

Themes from Six U.S. Air Force Major Defense Acquisition Programs

Mark A. Lorell, Robert S. Leonard, Abby Doll

RAND Project AIR FORCE

Prepared for the United States Air Force
Approved for public release; distribution unlimited

For more information on this publication, visit www.rand.org/t/RR630

Library of Congress Cataloging-in-Publication Data is available for this publication.

ISBN: 978-0-8330-8855-0

Published by the RAND Corporation, Santa Monica, Calif.

Preface

This report identifies and characterizes conditions present in six U.S. Air Force Major Defense Acquisition Programs (MDAPs) experiencing extreme cost growth, using case study analysis. This analysis draws on the findings of a companion document by Robert S. Leonard and Akilah Wallace, *Air Force Major Defense Acquisition Program Cost Growth Is Driven by Three Space Programs and the F-35A: Fiscal Year 2013 President's Budget Selected Acquisition Reports* (RAND Corporation, RR-477-AF, 2014), which provides detailed quantitative analysis of recent cost growth on Air Force MDAPs, including most of the six examined here. The case study analysis provided in this document is based on government program documentation and publically available open source materials, as well as interviews with program officials and subject matter experts. The purpose of this work is to characterize some of the key common attributes among the six programs with extreme cost growth. The end goal is to provide analysis that ultimately can be used to assist the Department of Defense and the Air Force in developing broad measures to improve cost and schedule outcomes for Air Force MDAPs.

The research reported here was commissioned by the Deputy Assistant Secretary for Acquisition Integration (SAF/AQX), Office of the Assistant Secretary of the Air Force for Acquisition, and conducted within the Resource Management Program of RAND Project AIR FORCE. The project monitor was the Technical Director of the Air Force Cost Analysis Agency.

RAND Project AIR FORCE

RAND Project AIR FORCE (PAF), a division of the RAND Corporation, is the U.S. Air Force's federally funded research and development center for studies and analyses. PAF provides the Air Force with independent analyses of policy alternatives affecting the development, employment, combat readiness, and support of current and future air, space, and cyber forces. Research is conducted in four programs: Force Modernization and Employment; Manpower, Personnel, and Training; Resource Management; and Strategy and Doctrine. The research reported here was prepared under contract FA7014-06-C-0001.

Additional information about PAF is available on our website:

http://www.rand.org/paf/

This report documents work originally shared with the U.S. Air Force on September 20, 2012. The draft report, issued on February 18, 2014, was scrutinized by formal peer reviewers and U.S. Air Force subject-matter experts.

Contents

Preface iii
Figures vii
Tables ix
Summary xi
Acknowledgments xvii
Abbreviations xix
1. Introduction 1
  Overview and Research Objectives 1
  Approach and Methodology 3
2. Case Studies 9
  Advanced Extremely High Frequency Satellite System 9
    Summary Overview of AEHF 9
    AEHF Case History 10
    AEHF Summary Findings of Key Program Characteristics 13
  C-130 Avionics Modernization Program (AMP) 13
    Summary Overview of C-130 AMP 13
    C-130 AMP Case History 15
    C-130 AMP Summary Findings on Key Program Characteristics 16
  Evolved Expendable Launch Vehicle 17
    Summary Overview of EELV 17
    EELV Case History 18
    EELV Summary Findings on Key Program Characteristics 21
  Global Hawk 22
    Summary Overview of Global Hawk 22
    Global Hawk Case History 23
    Global Hawk Summary of Key Program Characteristics 26
  National Polar-Orbiting Operational Environmental Satellite System (NPOESS) 27
    Summary Overview of NPOESS 27
    NPOESS Case History 29
    NPOESS Summary of Key Program Characteristics 31
  Space-Based Infrared Systems High 31
    Summary Overview of SBIRS High 31
    SBIRS High Case History 32
    SBIRS Summary of Key Program Characteristics 36
3. Summary Findings and Observations 37
  Summary Overview of the Six Programs 37

Key Lessons Learned and Their Relevance to Today 39
Recommendation #1: Ensure That Programs Have Realistic Cost Estimates at MS B 41
Recommendation #2: Embrace Evolutionary Strategies with Comprehensive and Proven Implementation Strategies 42
References 45

Figures

Figure 2.1. Advanced Extremely High Frequency Satellite 9
Figure 2.2. C-130 AMP Cockpit Simulator/Trainer Showing Cockpit Upgrades 14
Figure 2.3. The Boeing Delta IV Heavy and the Lockheed Atlas V Rocket Launchers 17
Figure 2.4. RQ-4B Block 40 Global Hawk 23
Figure 2.5. Artist's Rendition of the NPOESS Satellite in Orbit 27
Figure 2.6. The Number Two SBIRS GEO Satellite Undergoing Ground Testing 32

Tables

Table S.1. Two Categories of Common Characteristics of Six MDAPs with Extreme Cost Growth xiv
Table 1.1. Six Air Force MDAPs with Extreme Cost Growth 5
Table 3.1. Two Categories of Common Characteristics of Six MDAPs with Extreme Cost Growth 37

Summary

RAND recently analyzed cost growth trends in current U.S. Air Force major defense acquisition programs (MDAPs) using Selected Acquisition Report (SAR) data.[1] As a companion to that analysis, this report identifies and characterizes conditions present in six recent Air Force MDAPs that experienced extreme cost growth.[2] It is intended to supplement the quantitative analysis of 36 programs from the RAND SAR database reported in the companion document with a deeper detailed case-study examination of six of the most poorly performing programs to add specific program circumstances to provide a richer context for the quantitative analysis. This research was commissioned by the Deputy Assistant Secretary for Acquisition Integration, Office of the Assistant Secretary of the Air Force for Acquisition.[3]

It is important to note that these programs are not typical or representative Air Force programs. Rather, they are the "worst of the worst" in terms of high cost growth and schedule slippage. Since these types of programs cause the most budgetary and programmatic disruption to the entire Air Force portfolio of development programs, we felt it should be our highest priority to assist the Air Force in avoiding these types of extreme outcomes in the future. And they are not complete outliers. As our companion document shows, the typical recent Air Force MDAP experiences substantial cost growth. It is therefore important to identify the characteristics and conditions of the worst of the worst-performing acquisition programs. We believe this in-depth qualitative case study analysis, read in conjunction with the quantitative analysis of 36 programs provided in the companion document, assists in providing a more nuanced and sophisticated understanding of current MDAP acquisition trends.

It is important to note, however, that while this approach identifies common characteristics and conditions of poorly performing programs, it is not sufficient for definitively identifying the true causes of extreme cost growth and their relative importance. To achieve that objective, it is necessary to compare the worst-performing programs to an equal and similar set of "control" programs that have performed relatively well. The mere fact that all six poorly performing programs share a common attribute is meaningless if it is found that the same attribute is shared by relatively well-performing programs. Thus, the characterization of the attributes of the six programs in this report is only the first step of our analysis. Research currently underway by

[1] See Robert S. Leonard and Akilah Wallace, *Air Force Major Defense Acquisition Program Cost Growth Is Driven by Three Space a Program and the F-35: Fiscal Year 2013 President's Budget Selected Acquisition Reports*, Santa Monica, Calif.: RAND Corporation, RR-477-AF, 2014.

[2] We define extreme cost growth as a percentage cost growth which is at least one standard deviation above the mean cost growth of all programs in one of five cost growth categories used for measuring cost growth for Air Force MDAPs, as shown in Table 1, p. 4.

[3] The Project Monitor was the Technical Director, Air Force Cost Analysis Agency.

RAND Project AIR FORCE's Resource Management Program aims, among other things, at providing the analysis of the "control" set of better-performing programs to determine whether the attributes common to the worst-performing programs are the true drivers behind extreme cost growth.[4] The purpose of the research reported here is to characterize the key conditions and attributes common to the six programs that experienced extreme cost growth. This completed research, combined with current ongoing research, is intended ultimately to provide analysis that will assist the Air Force in developing improved acquisition policies and procedures that will contribute to better program outcomes in the areas of cost, schedule, and performance and help reduce the likelihood of future programs experiencing extreme cost growth. This document draws on a variety of sources, including government program documentation such as SAR program descriptions and histories and analysis included in the program Defense Acquisition Executive Summaries (DAES), as well as publically available open source materials. On occasion, this information was supplemented by interviews with current or former program officials or other subject matter experts.

To achieve the necessary fine-grained knowledge and insight into specific programs, we limited the number of programs to the worst-performing Air Force MDAPs in terms of cost growth (all of which experienced "extreme cost growth" as we define it). There were six programs meeting our criteria, listed here in alphabetical order:

- Advanced Extremely High Frequency (AEHF) satellite system
- C-130 Avionics Modernization Program (AMP)
- Evolved Expendable Launch Vehicle (EELV) program
- Global Hawk (RQ-4 Global Hawk high-altitude long-endurance [HALE] unmanned aerial vehicle)
- National Polar-Orbiting Operational Environmental Satellite System (NPOESS)
- Space-Based Infrared System High (SBIRS High).

Our in-depth qualitative analysis of their programmatic histories indicates that two main categories of common characteristics and conditions, comprised of five sub-elements, were prominent in these programs:[5]

- premature approval of Milestone (MS) B
  - insufficient technology maturity and high integration complexity
  - unclear, unstable, or unrealistic requirements
  - unrealistic cost estimates
- suboptimal acquisition strategies and program structure

---

[4] For an overview of statistical approaches to causal analysis, see Guido W. Imbens and Donald B. Rubin, *Causal Inference in Statistics, Social, and Biomedical Sciences*, Cambridge University Press, 2015.

[5] The U.S. Government Accountability Office developed a similar finding based on the assessment of 54 programs in 2005. See U.S. Government Accountability Office, *Defense Acquisitions: Assessments of Selected Major Weapon Programs*, Washington, D.C., GAO-05-301, March 2005.

- adoption of acquisition strategies and program structures that lacked adequate processes for managing risk through incrementalism and provision of appropriate oversight and incentives for the prime contractor
- use of a combined MS B/C milestone or premature award of MS C prior to the achievement of adequate production article design stability.

These characteristics and conditions are summarized in Table S.1.[6]

The most significant characteristic, which spanned all six programs, was premature approval of MS B. Our research indicates that none of these programs was ready for MS B approval, usually for multiple reasons. Five programs were characterized by immature designs and technology or failure to recognize the complexity of system integration, combined with insufficient programmatic and technological risk reduction efforts. These programs also suffered from unstable requirements that were incomplete, unclear, or disputed. Perhaps most striking, every one of the six programs suffered from serious cost-estimation issues. Nearly all the programs failed to place sufficient emphasis on the actual costs of similar or related predecessor programs. Five of the six programs passed MS B with at least some of the stakeholders or other interested parties aware that the MS B baseline cost estimates were unrealistic. In addition, most of the programs began with unclear, unstable, or unrealistic performance requirements and expectations and did not include an institutionalized process for modulating requirements in the interest of affordability as the program progressed. This issue is closely related to the category of immature technologies and integration complexity. The lack of realism in the cost estimates was also closely linked to the prior two elements, in that the difficulty and complexity of the required developmental and production efforts were underestimated, as were the effects of unrealistic or unstable requirements.

A second significant characteristic common across all six programs was the use of inappropriate acquisition strategies or program structures. None of the programs systematically implemented evolutionary acquisition strategies as a tool to help manage higher-risk technology and system integration programs. Several programs emphasized an acquisition approach employing a single step to full capability, which increases risk. This is completely acceptable, particularly if there is an urgent need for the capability, but acquisition and budget officials must be fully cognizant of the possible cost implications.

Another key element of this characteristic in many of the programs was the use of unproven acquisition strategies that, at least in the way they were implemented, failed to encourage adequate Air Force oversight of the program and the prime contractor and encouraged optimistic cost savings estimates through the planned extensive use of commercial-off-the-shelf (COTS) technologies, civil-military integration (CMI), and commercial-type contracting and management approaches. These strategies emerged from the acquisition reform initiatives and legislation launched in the early to mid-1990s in a cost-constrained environment similar in some respects to

[6] As shown in Table S.1, not all programs examined experienced all the characteristics and conditions identified.

the current one. Therefore, it is particularly important to understand the characteristics and roles played in the past by new strategies, technologies, and processes promising large savings. Finally, a related important characteristic was the failure to ensure that the design was sufficiently stable and the technology adequately matured prior to approval of MS C and low-rate initial production (LRIP). This problem is clearly linked to the simultaneous or near simultaneous approval of MS B and MS C. Approval of a combined MS B/C is risky unless the item to be procured is already adequately mature with a stable production design and thus fully ready for production. Otherwise, discovery of design flaws and other technical issues during continuing development beyond MS C can lead to the need to make costly modifications and retrofits to already produced production articles.

Table S.1. Two Categories of Common Characteristics of Six MDAPs with Extreme Cost Growth

	AEHF	C-130 AMP	EELV[a]	Global Hawk	NPOESS	SBIRS High
Premature MS B						
Immature technology; integration complexity	√	√		√	√	√
Unclear, unstable, or unrealistic requirements	√	√		√	√	√
Unrealistic cost estimates	√	√	√	√	√	√
Acquisition policy and program structure						
Inappropriate acquisition strategy and program structure	√	√	√	√	√	√
MS B/C (premature MS C)	√		√	√	√	
Unit total (PAUC) cost growth	95%	193%	273%	152%	154%	279%

NOTES: The bottom line of table shows Program Average Unit Cost (PAUC) cost growth for each program. Note there is little or no correlation between the number of characteristics evident in a specific program and the severity of that programs cost growth in percentage terms. Each MDAP is unique in context and circumstances, and this is why it is so important to convey the details of each case history, and ultimately to compare these six "worst of the worst" cases to better-performing MDAPs.

[a] The EELV acquisition strategy may not have been inappropriate given reasonable assumptions held at the beginning of the program, but these assumptions proved optimistic and ultimately incorrect.

Based on our analysis of the key program characteristics and conditions in these six MDAP case studies as summarized in Table S.1, we arrived at two broad categories of potential recommendations:

- MDAPs must have credible baseline cost estimates at MS B to provide realistic baseline metrics for accurately measuring real cost growth.
- The Air Force should develop, refine, and implement robust evolutionary or incremental acquisition strategies and policies that reduce and control technological and programmatic risk, unless timely operational need has clear priority over cost savings.

Under the rubric of the first broad recommendation, we urge the Air Force to consider the following broad measures:

- Recognize and incorporate the strong predictive relevance of predecessor programs' costs when establishing new program baseline cost estimates and budgets at MS B.
- Maintain a healthy skepticism toward claims that new approaches or technologies will substantially reduce costs of future systems compared with past systems, especially when greater capabilities are promised or desired.
- Address credible alternative cost and risk assessments from the Office of the Secretary of Defense, the U.S. Government Accountability Office, the Congressional Budget Office, the Congressional Research Service, and other authoritative sources prior to and at the time of MS B.

Regarding the second broad recommendation, we urge the Air Force to embrace, refine, and implement appropriate evolutionary or incremental acquisition strategies encompassing the following broad guidelines, most of which apply to the pre-MS B period, as shown below. When this is not possible or desirable due to operational requirements and threat assessments, it is important to recognize and prepare for the likely higher probability of substantial cost growth.

- Adopt revolutionary technologies only when necessary, such as when required to counter relatively near-term threats. Reemphasize evolutionary acquisition strategies to achieve full objective capabilities through a series of separate lower-risk program increments or steps.
- Begin a new Acquisition Category (ACAT) I MDAP only when objective capabilities and goals cannot reasonably be met through a series of smaller, less risky ACAT II-IV programs.
- Conduct early and comprehensive cost-benefit and risk assessments of requirements and technology, as well as system design and integration, all with an emphasis on affordability.
- Reach consensus on requirements and costs among all stakeholders before determining final formal requirements.
- Ensure adequate oversight of the prime contractor but use positive incentives to motivate contractors to pursue affordability and cost-saving initiatives.
- Minimize overlap within and among specific evolutionary program increments, as well as between major overall program phases. Avoid granting simultaneous MS B/C approval, unless the item is fully production ready. Ensure design stability and maturity prior to the launch of LRIP to avoid the need for costly retrofits on production items based on continuing research, development, test, and evaluation (RDT&E) activities.

Acknowledgments

This work would not have been possible without the sustained research commissions of Mr. Rich Lombardi, Deputy Assistant Secretary for Acquisition Integration (SAF/AQX; and Mr. Blaise Durante before him) and Ms. Ranae Woods (and before her Mr. Jay Jordon), Technical Director of the Air Force Cost Analysis Agency.

The support and encouragement of Laura Baldwin, former director of the RAND Project AIR FORCE (PAF) Resource Management Program, and Bill Shelton, Cost Umbrella Project Leader, have been crucial to the pursuit of this analysis. The insights provided by Lara Schmidt, PAF associate director and research quality assurance manager, greatly strengthened this document.

The authors would also like to thank Jack Graser for his guidance over the years, and Obaid Younossi before him. We are also grateful to Fred Timson for his mentoring over the past two decades, and Jeff Drezner for the SAR data we inherited from him in the mid-1990s.

The authors benefited greatly from the constructive suggestions and useful insights provided by our two formal reviewers, Edward Keating and M. J. Hicks. We thank them for their efforts.

Abbreviations

ACAT	acquisition category
ACTD	Advanced Concept Technology Demonstration
AEHF	Advanced Extremely High Frequency satellite system
AFCAA	Air Force Cost Analysis Agency
ALARM	Alert Locate and Report Missiles
AMP	Avionics Modernization Program
AWS	Advanced Warning System
CAAP	Common Avionics Architecture for Penetration
CAIG	Cost Analysis Improvement Group
CAPP	Common Avionics Architecture for Penetration
CBO	Congressional Budget Office
CMI	civil-military integration
CMIS	Conical-scanning Microwave Imager/Sounder
COTS	commercial off-the-shelf
CrIS	Cross-Track Infrared Sounder
DAES	Defense Acquisition Executive Summaries
DARPA	Defense Advanced Research Projects Agency
DMSP	Defense Meteorological Satellite Program
DoD	U.S. Department of Defense
DoDI	Department of Defense Instruction
DSP	Defense Support Program
EELV	Evolved Expendable Launch Vehicle
EMD	engineering and manufacturing development
FAR	Federal Acquisition Regulation
FEWS	Follow-on Early Warning System
FOC	full operational capability

FY	fiscal year
GAO	U.S. Government Accountability Office; formerly the U.S. General Accounting Office
GATM	Global Air Traffic Management
GEO	geosynchronous earth orbit
GOTS	government off-the-shelf
HALE	high altitude long endurance
HEO	highly elliptical orbit
IC	Intelligence Community
IPO	integrated program office
ISR	intelligence, surveillance, and reconnaissance
JROC	Joint Requirements Oversight Council
MDAP	Major Defense Acquisition Program
Milstar	Military Strategic and Tactical Relay
MS	milestone
MUA	military utility assessment
NASA	National Aeronautics and Space Administration
NOAA	National Oceanic and Atmospheric Administration
NPOESS	National Polar-Orbiting Operational Environmental Satellite System
OMB	Office of Management and Budget
OMPS	Ozone Mapping and Profiler Suite
OSD	Office of the Secretary of Defense
OTA	Other Transactions Authority
PAF	RAND Project AIR FORCE
PAUC	program average unit cost
PBA	price-based acquisition
POES	Polar Operational Environmental Satellite
RDT&E	research, development, test, and evaluation
RFP	request for proposal

SAF/AQ	Office of the Assistant Secretary of the Air Force for Acquisition
SAR	Selected Acquisition Report
SBIRS	Space-Based Infrared System
SDD	system development and demonstration
SME	subject matter expert
SSPR	Shared System Performance Responsibility
SV	space vehicle
TCS	transformational communications study
TRD	technical requirements document
TSAT	Transformational Satellite Communications System
TSPR	Total System Program Responsibility
TY	then year
UAS	unmanned aircraft system
ULA	United Launch Alliance
USSOCOM	U.S. Special Operations Command
VIIRS	Visible/Infrared Imager Radiometer Suite
WSARA	Weapon System Acquisition Reform Act of 2009
WR-ALC	Warner Robins Air Logistics Center

1. Introduction

Overview and Research Objectives

RAND recently analyzed cost growth trends in current U.S. Air Force Major Defense Acquisition Programs (MDAPs) using Selected Acquisition Report (SAR) data.[1] As a companion to that analysis, this report identifies and characterizes conditions present in six recent Air Force MDAPs that experienced extreme cost growth.[2] It is intended to supplement the quantitative analysis of 36 programs from the RAND SAR database reported in the companion document with a deeper detailed case-study examination of six of the most poorly performing programs to add specific program circumstances to provide a richer context for the quantitative analysis. This research was commissioned by the Deputy Assistant Secretary for Acquisition Integration, Office of the Assistant Secretary of the Air Force for Acquisition.[3]

It is important to note that these programs are not typical or representative Air Force programs. Rather, they are the "worst of the worst" in terms of high cost growth and schedule slippage. Since these types of programs cause the most budgetary and programmatic disruption to the entire Air Force portfolio of development programs, we felt it should be our highest priority to assist the Air Force in avoiding these types of extreme outcomes in the future. And they are not complete outliers. As our companion document shows, the typical recent Air Force MDAP experiences substantial cost growth. It is therefore important to identify the characteristics and conditions of the worst of the worst-performing acquisition programs in terms of cost growth. We believe this in-depth qualitative case study analysis, read in conjunction with the quantitative analysis of 36 programs provided in the companion document, assists in providing a more nuanced and sophisticated understanding of current MDAP acquisition trends.

It is important to note, however, that while this approach identifies common characteristics and conditions of poorly performing programs, it is not sufficient for definitively identifying the true causes of extreme cost growth and their relative importance. To achieve that objective, it is necessary to compare the worst-performing programs with an equal and similar set of "control" programs that have performed relatively well. The mere fact that all six poorly performing programs share a common attribute is meaningless if it is found that the same attribute is shared

[1] See Robert S. Leonard and Akilah Wallace, *Air Force Major Defense Acquisition Program Cost Growth Is Driven by Three Space a Program and the F-35: Fiscal Year 2013 President's Budget Selected Acquisition Reports*, Santa Monica, Calif.: RAND Corporation, RR-477-AF, 2014.

[2] We define extreme cost growth as a percentage cost growth that is at least one standard deviation above the mean cost growth of all programs in one of five cost growth categories used for measuring cost growth for Air Force MDAPs, as shown in Table 1.1 later in this chapter.

[3] The project monitor was the Technical Director, Air Force Cost Analysis Agency.

by relatively well-performing programs. Thus, the characterization of the attributes of the six programs in this report is only the first step of our analysis. Research currently underway by RAND Project AIR FORCE's Resource Management Program aims, among other things, at providing the analysis of the "control" set of better-performing programs to determine whether the attributes common to the worst-performing programs are the true drivers behind extreme cost growth.[4]

To best identify and characterize common attributes of all six poorly performing programs, the RAND research team produced detailed analytical program histories based on program documentation and other available U.S. government and open source information, which built on the detailed cost analyses based on data from SARs found in the project's companion document.[5] This permitted a thorough assessment of the specific circumstances and history of each MDAP via in-depth case studies. The government information we used consisted of SAR program descriptions and histories, and events and analyses included in the program Defense Acquisition Executive Summaries (DAES). We also used a comprehensive selection of open published and unpublished literature. On occasion this information was supplemented by interviews with current or former program officials, or other subject matter experts (SMEs).[6]

The need to develop a more precise and refined understanding of the common characteristics of poorly performing programs, and ultimately the true root causes of MDAP cost growth, is particularly crucial during the second decade of the 21st century. The United States is entering a sustained period in which downward pressures on the Department of Defense (DoD) acquisition budget are likely to become increasingly intense and impose new constraints. Budgetary resources and flexibility are likely to continue to decline. The 1990s witnessed a similar period of sustained acquisition budgetary decline, which, like the present period, led to the formulation of many new and often unproven acquisition approaches that promised to increase efficiency and reduce costs. Several of these strategies, at least in the way they were implemented, failed to encourage adequate Air Force oversight of the program and the prime contractor and encouraged optimistic cost savings estimates through the planned extensive use of commercial-off-the-shelf (COTS) technologies, civil-military integration (CMI), and commercial-type contracting and management approaches. Many of these strategies are identified with programs suffering from extreme cost growth and include many of the case studies examined in this report. In the present similar environment of constrained budgets, it is particularly important not to repeat the implementation errors of the past regarding new strategies and technologies and processes promising unrealistically large savings. This is the central goal of this analysis.

---

[4] For an overview of statistical approaches to causal analysis, see Guido W. Imbens and Donald B. Rubin, *Causal Inference in Statistics, Social, and Biomedical Sciences*, Cambridge University Press, 2015.

[5] Leonard and Wallace, 2014.

[6] Many of these interviews took place in the spring of 2006 in support of an earlier RAND project.

Approach and Methodology

To achieve the project objective, we built on earlier RAND work in the area of program cost growth. Two earlier RAND reports document these efforts: *Sources of Weapon System Cost Growth*, published in 2008, and *Historical Cost Growth of Completed Weapon System Programs*, published in 2006.[7]

Based on these prior analyses, the project team concluded that meaningful insights into the root causes of MDAP cost growth required developing in-depth qualitative case studies to supplement the statistical analysis derived from the SAR data. To achieve the depth of detail needed to attain insight into specific programs, the study team limited MDAP case studies to the worst-performing MDAPs (in terms of cost growth from Milestone [MS] B) over the past several years, which are of direct interest and relevance to the Air Force. Narrowing the total number of cases examined permitted the collection of the high level of detail and insight into the structure and implementation of the programs necessary to ultimately understand the dynamics behind the cost growth that took place.

As noted above, we do not compare our six worse cases with other MDAP cases with less cost growth in this report. Thus, we do not claim that the common characteristics we identified of the poorly performing programs are necessarily the key causes of extreme cost growth. We will be able to fully understand the key cost drivers and causes of extreme cost growth only after we complete our current research on programs that experienced relatively positive outcomes. Nonetheless, it is still important to understand the key program characteristics of programs that are the "worst of the worst." First of all, the overall dataset of 36 programs examined in the quantitative analysis included in the companion document shows that most MDAPs experience substantial cost growth. Only five of the 36 programs analyzed experienced no cost growth. All but one of the seven continuing programs examined experienced greater than 50 percent cost growth.[8] Thus, significant cost growth is the norm for continuing MDAPs, and it is likely that the principle causes of cost growth in most programs are similar. Second, the programs with extreme cost growth have caused the most disruption to Air Force acquisition budgeting and planning, and have led often to program cancellations, which have at best greatly delayed the ability of the Air Force to meet important combatant requirements in a timely manner. Programs with extreme cost growth are disproportionately damaging to the overall Air Force portfolio planning process and recapitalization objectives, and thus should be a key focus of any attempts to understand the root causes of the problem.

---

[7] Joseph G. Bolten, Robert S. Leonard, Mark V. Arena, Obaid Younossi, and Jerry M. Sollinger, *Sources of Weapon System Cost Growth: Analysis of 35 Major Defense Acquisition Programs*, Santa Monica, Calif.: RAND Corporation, MG-670-AF, 2008; Mark V. Arena, Robert S. Leonard, Sheila E. Murray, and Obaid Younossi, *Historical Cost Growth of Completed Weapon System Programs*, Santa Monica, Calif.: RAND Corporation, TR-343-AF, 2006.

[8] As measured in terms of Program Acquisition Unit Cost (PAUC) growth. See Figure 1.4, Leonard and Wallace, 2014.

Our selection criterion for the worst-performing recent Air Force MDAPs was straightforward. Based on past analyses of MDAP cost growth using hundreds of programs dating back over four decades, RAND cost analysts defined the programs with the worst cost growth as those with a percentage cost growth at least one standard deviation above the mean cost growth percentage of the total sample. These programs are described as exhibiting "extreme cost growth." This project's companion document, which analyzed the most recent cost growth trends among MDAPs of interest to the Air Force and currently reporting SARs over the past several years, included approximately 30 MDAPs.[9] Of these, six programs, or approximately 20 percent, experienced extreme cost growth according to our definition.[10] These programs are listed below in alphabetical order:

- Advanced Extremely High Frequency (AEHF) satellite system
- C-130 Avionics Modernization Program (AMP)
- Evolved Expendable Launch Vehicle (EELV) program
- Global Hawk (RQ-4 Global Hawk high-altitude long-endurance [HALE] unmanned aerial vehicle)
- National Polar-Orbiting Operational Environmental Satellite System (NPOESS)
- Space-Based Infrared System High (SBIRS High).

As shown in Table 1.1, all six programs experienced extreme cost growth in at least two of the standard five metrics used in RAND's cost growth analysis. These are grouped under the two broad categories of budgetary and unit cost growth. There are three types of budgetary cost growth: development, procurement, and program.[11] There are also two types of unit cost growth: unit procurement and unit program.[12] The broad budget category has not been adjusted for quantity changes, so it shows the budgetary effects in constant dollars of cost growth. The last two columns on the right measure unit cost growth from the original MS B estimate and are adjusted for final program quantities, thus providing the best assessment of the estimate's accuracy.[13] Of the twelve unit measurements, ten are extreme (as shown in bold type font in Table 1), indicating gross underestimates of these programs at MS B.

---

[9] Not all these programs are included in the full quantitative analysis in the companion document. That was because programs that were cancelled (or had major segments eliminated) during the period under consideration were excluded from the quantitative analysis because they didn't have full time series of data. See Leonard and Wallace, 2014.

[10] Leonard and Wallace, 2014. Historically approximately 10 percent of the total programs in the database showed extreme cost growth. The higher percentage with extreme cost growth among the current programs does not necessarily indicate that outcomes of more recent programs are worse than historical programs, due to a variety of technical cost comparison issues. For further discussion, see Leonard and Wallace, 2014, and footnote 11 above.

[11] Program costs are defined as the sum of development, procurement, military construction, and acquisition related operations and maintenance costs associated with each program's acquisition.

[12] Unit program cost is defined as total program cost (development, procurement, military construction, and acquisition related operations and maintenance)) divided by total adjusted program units.

[13] Department of Defense Instruction (DoDI) 5000.2, *Operation of the Defense Acquisition System*, May 12, 2003, designates MS B as the formal beginning of an MDAP.

All six of these programs experienced extreme cost growth that, according to our analysis of the SARs, is not explicable due to substantial quantity increases or unforeseeable circumstances taking place beyond MS B which were outside of the program's control. How can the extreme cost growth of these six MDAPs be explained? What are the true sources and most important drivers of cost growth in these programs? To answer these questions, it is necessary to construct a much more detailed case history of each of them than was possible based solely on the narrative and quantitative data in SARs. In forthcoming reports, we will compare the outcome of this study to the characteristics of better performing programs to better assist us in identifying the key causal factors in extreme cost growth.

We used three major sources of data and information to assemble our case histories and qualitative analyses. As noted earlier, almost all cost data and cost growth analysis was derived from the SARs. The SARs also include annual program summary histories and many other details of the programs useful for developing case histories.

Table 1.1. Six Air Force MDAPs with Extreme Cost Growth

		Budgetary Cost Growth			FY 2012	Unit Cost Growth	
Program	MS B or B/C	Development	Procurement	Program	M$ Growth	Procurement	Program
AEHF	November 2001	58%	**325%**	119%	$7,600	**217%**	95%
C-130 AMP*	July 2001	148%	24%	47%	$2,000	**194%**	**193%**
EELV	October 1998	29%	**229%**	**210%**	$36,700	**299%**	**273%**
Global Hawk*	March 2001	**277%**	123%	157%	$8,800	86%	**152%**
NPOESS*	August 2002	106%	101%	68%	$4,800	**335%**	**154%**
SBIRS High	November 1996	**235%**	**574%**	**315%**	$14,800	**407%**	**279%**

NOTES: Percentages shown in bold represent extreme cost growth, defined as cost growth more than one standard deviation above the mean for that measure. Programs listed with an asterisk were terminated or truncated in the FY 2013 President's Budget.

To provide a more robust foundation for our case studies, we reviewed the DAES for each of these programs. Throughout the life of the program, these reports were typically submitted at least four times per year and, during certain years, as many as ten times per year.[14] DAES often provide more and different types information than the SARs regarding program issues and events, since, unlike SARs, they are generated only for internal DoD use. They also include comments and insights from multiple sources within the Air Force as well as within other components of DoD.

Finally, we conducted wide-ranging surveys of open source information on these programs, including government studies such as those from the U.S. Government Accountability Office

[14] DAES usually begin being issued following a MS A or equivalent decision, but typically include very sparse data and information until the MS B decision time period.

(GAO)[15] and Congressional Budget Office (CBO), studies and dissertations from a variety of DoD sources and non-DoD sources, and the extensive material available in the trade press, such as *Inside Defense* and *Aviation Week and Space Technology*. This third set of source material from outside the official program documentation proved to be crucial for establishing the broader context regarding many key issues of relevance to our inquiry. In addition, in some instances we were able to conduct interviews with current or former program officials or other SMEs. We also reviewed past documented interviews with senior program officials conducted in support of earlier RAND projects.

Based on our assessment of these six programs, we found that each program exhibited major shortcomings in as many as five sub-element areas, which we placed under two broader categories:

- premature approval of MS B
  - insufficient technology maturity and high integration complexity
  - unclear, unstable, or unrealistic requirements
  - unrealistic cost estimates
- suboptimal acquisition strategies and program structure
  - adoption of acquisition strategies and program structures which lacked adequate processes for managing risk through incrementalism and provision of appropriate oversight and incentives for the prime contractor
  - use of a combined MS B/C milestone or premature award of MS C prior to the achievement of adequate production article design stability.

It is important to note that each of the three sub-elements under premature approval of MS B may have made later cost growth highly likely. In other words, our research may suggest that in these six program case studies, the likelihood of extreme cost growth was virtually predetermined at MS B. Thus, in the future, the remedies to these sorts of problems must be sought in the pre–MS B phases of programs. However, the final determination must await the outcome of our analysis and comparison to programs that performed relatively well.

Most of the programs began with inadequately matured technologies, underestimations of integration complexity, and unrealistic cost and performance expectations or unstable requirements, and they did not include an institutionalized approach to modulate requirements in the interest of affordability as the program progressed. A key insight from this analysis is that in almost every case, much less optimistic independent cost estimates were available, and, in many cases, widely known among program officials. This raises an important question: Why were the more unrealistic cost estimates selected for the program baseline when other, more reasonable alternative estimates were available? Obviously a variety of political, budgetary, and programmatic issues encouraged programs to select and formalize the lower, less realistic cost

[15] Formerly the U.S. General Accounting Office.

estimates. This suggests that the acquisition process is skewed and incentivizes the wrong behavior.[16]

Nearly as important was the failure to adopt appropriate acquisition strategies and program structures, particularly evolutionary strategies, as tools to help manage higher-risk technology programs. Several programs emphasized an acquisition approach employing a single step to full capability, which increases risk. Another related major characteristic associated with four of the programs with extreme cost growth was program launch with a combined MS B/C. Such an approach can indicate extreme concurrency between development and production phases, and is only appropriate when the item being procured has a stable and mature design at the combined milestone. When major development and testing activities continue into the production phase, the item may not be truly ready for production, leading to the need for expensive modifications and retrofits later in the program.

It is important to emphasize that a key characteristic associated with programs experiencing extreme cost growth is the use of inappropriate or unproven acquisition strategies. At least five of the six used acquisition strategies that, at least in the way they were implemented, failed to encourage adequate Air Force oversight of the program and the prime contractor, and raised false expectations regarding untested technological and programmatic approaches to cost savings. These strategies emerged from the acquisition reform initiatives and legislation launched in the early to mid-1990s. Some of these measures—such as various forms of the total system program responsibility (TSPR) concept,[17] exploitation of COTS technical and design solutions, and commercial-like contracting approaches such as price-based acquisition (PBA)—contributed to decreased government planning and oversight of programs, and to unrealistic cost savings expectations. These acquisition reform measures of the 1990s were introduced primarily under the belief that they would reduce program costs, which was necessary due to dramatic declines in the defense budget following the end of the Cold War. It is therefore particularly important now, as DoD enters a new era of increasingly constrained defense budgets and officials seek new and sometimes less familiar acquisition approaches with the intent to reduce program costs, that the key characteristics of programs experiencing extreme cost growth that were launched in the 1990s during an analogous period of declining budgets are fully comprehended and not repeated.

We present additional discussion of our findings from the six case studies after our discussion of the program case studies. The next chapter presents an in-depth discussion of each of our case studies of these six programs, aimed at identifying their major common characteristics. The cases are presented in alphabetical order.

---

[16] The Weapons Systems Acquisition Reform Act of 2009 (WSARA) legislated many reforms, some of which were aimed at addressing issues touched on in this paragraph. WSARA requires independent cost estimates and certification of full funding to those cost estimates.

[17] TSPR was actually first formulated under Secretary of Defense Robert MacNamara in the early 1960s and applied to the TFX/F-111 program. While not always called TSPR in the 1990s, essentially the same concept was applied to many programs during the Clinton administration acquisition reform period.

2. Case Studies[1]

Advanced Extremely High Frequency Satellite System

Summary Overview of AEHF

The Advanced Extremely High Frequency Satellite System (AEHF, shown in Figure 2.1) is a joint multi-service satellite communications system currently managed by the Air Force, providing secure and jam-resistant global communications between the highest leadership levels in the United States and the land, sea, and air forces during peacetime and conflict, up to and including nuclear war. AEHF is a follow-on to the 1990s-era Military Strategic and Tactical Relay (Milstar) Block II satellites and system.

Figure 2.1. Advanced Extremely High Frequency Satellite

SOURCE: Image shared by Lockheed Martin Space Systems Company, via Flickr; no known copyright restrictions.

[1] All the specific factual information reported in these brief case history summaries is widely available from a variety of standard aerospace industry trade publications, program SARs, and other open sources. However, our interpretations of the information are informed by a careful reading of the DAES, as well as, in many cases, interviews with program officials and other SMEs, often conducted in the course of earlier RAND Project AIR FORCE research projects.

AEHF is composed of three segments: four satellites in geosynchronous earth orbit (GEO) providing worldwide coverage, a mission control segment, and a terminal segment. The mission control segment controls and monitors the satellites and is designed to be highly survivable, including both fixed and mobile stations. The terminal segment includes both fixed and mobile components, including terminals on land, ships, submarines, and aircraft. Allied countries, including the United Kingdom, Canada, and the Netherlands, have signed agreements to buy terminals. All segments aim at a high level of security and survivability in all threat environments.

Lockheed Martin Space Systems Company, in Sunnyvale, California, is the AEHF prime contractor, space and ground segments provider, and system integrator. Northrop Grumman Aerospace Systems, in Redondo Beach, California, provides the actual satellite payload, under contract to Lockheed Martin.

Our analysis indicates that the most important AEHF program characteristics common with the other programs we examined that experienced extreme cost growth were unstable program requirements and schedule uncertainty, immature technology at MS B/C, unrealistic cost estimates at MS B/C, and unstable program structure. The uncertainty and instability surrounding requirements and schedule arose after the unanticipated failure of the launch of a Milstar Block 2 satellite in April 1999 led to conflicts among the AEHF stakeholders over how to adjust the AEHF schedule and requirements to quickly fill in for the lost capability. These conflicts were neither quickly nor fully resolved, and, when combined with issues regarding the application of immature technologies, they appear to be linked to cost growth experienced as the program progressed.

AEHF Case History

AEHF originally emerged in the early 1990s as a proposed Milstar 3 replenishment follow-on to Milstar 1 and 2, combined with ground mission control software upgrades. In January 1995, Milstar 3 was split off from the Milstar program and became a separate program renamed AEHF. The baseline Milstar program was then stabilized at two Block 1 satellites and four Block 2 satellites. As defense budgets continued their decline following the end of the Cold War, AEHF became a candidate for testing new and sometimes untried acquisition strategies intended to reduce costs while increasing efficiency and system capabilities. Indeed, from its very inception, acquisition officials asserted that the new acquisition approaches, combined with the maximum use of commercial technology, could produce AEHF space vehicles (SVs) with greater capabilities than the originally planned Milstar 3, yet with reduced satellite and launch costs due to decreased weight permitting the use of cheaper medium launch vehicles.

The initial AEHF acquisition strategy envisioned a pre-MS B[2] competitive phase between two contractor teams for concept definition and risk reduction. In August 1999, two 18-month contracts were awarded to a Lockheed/TRW team and a Hughes team, with a planned down select to the winning contractor team in early 2001. All three contractors had been major players on the Milstar program. Unfortunately, in April 1999 Milstar Flight 3 (a Milstar Block 2 satellite) failed to achieve the proper geosynchronous orbit, leading to major concerns regarding significant delays in planned required capabilities.

This led to an extended and ultimately unresolved debate among the Office of the Secretary of Defense (OSD), the Air Force, and contractor stakeholders over how AEHF schedule and capabilities could be realistically restructured, and at what cost, to compensate for the Milstar launch failure. OSD argued for shortening the development schedule by moving the launch date of the first AEHF up by 18 months, from mid-2006 to late 2004. The contractors and the Air Force opposed this aggressive schedule. The Air Force objected mainly on budgetary grounds, while the contractors were concerned about requirements clarity and technological complexity.[3]

The Joint Requirements Oversight Council (JROC) accepted the need for this truncated schedule in December 1999 and authorized examination of alternative approaches, but the debate continued over what overall schedule, cost, and capabilities were feasible. This continued throughout 2000 and 2001, with OSD insisting on the launch of a full-capability AEHF in December 2004. The contractors proposed a compromise of launching a significantly down-scoped, less capable "Pathfinder" AEHF according to the same shortened schedule; a full-capability AEHF would then later complete development and be made available several years later. The contractors maintained their position, since the technological complexity and performance requirements continued to grow, in part due to the multiple users involved in the program and their many requirements.

By the time the formal MS B/C decision was taken in October 2001, the various stakeholders had agreed to disagree and move ahead with the program, with fundamental issues remaining unresolved. There were four major elements to the compromise that permitted moving forward with the program. First, the competitive concept formulation and risk reduction phase was ended early, and all participating contractors were folded into a single "national team." While this was intended to consolidate resources and shorten schedule, it actually contributed to delays, as the formerly competing contractors struggled to combine their efforts. Second, the contractor and Air Force position was seemingly accepted, in that the first AEHF SV would be a full-capability

[2] In the 1990s, acquisition milestones were all labeled with numbers, whereas the current convention uses letters. For example, in the 1990s MS B was labeled MS II. In addition, a separate formal space acquisition document included slight differences in milestone designations for space programs compared with other programs. For the sake of simplicity, this document uses the current equivalent standard milestone designations using letters, as delineated in DoDI 5000.02, *Operation of the Defense Acquisition System*, December 8, 2008.

[3] RAND interview with AEHF program officials and other SMEs, January 24, 2006. Also see "AEHF (Advanced Extremely High Frequency) Series," *Jane's Space Systems and Industry*, June 14, 2012.

SV launched in June 2006, according to the original schedule prior to the Milstar launch failure. However, OSD compensated for this concession by requiring a two-year shortening of the overall AEHF schedule, moving full operation capability (FOC) of the overall system from fiscal year (FY) 2012 to FY 2010. This was to be accomplished by shortening the acquisition schedule of the third AEHF SV by two years. Finally, because of continually expanding performance requirements and persistent contractor concerns about requirements churn and growing technological complexity and risk, an acquisition decision memorandum at MS B established the Transformational Communications Study (TCS) to examine alternative architectures to achieve FOC by FY 2010 with three AEHFs and some combination of other systems or factors. The TCS rapidly morphed into a totally new space program, the Transformational Satellite Communications System (TSAT), which was meant to either supplement or replace AEHF at some unspecified time.

From this point through 2009, the failure to truly resolve the debate at MS B over capabilities, schedule, technology, and costs led to constant churn and uncertainty in the AEHF program over requirements, schedule, cost, program structure, procurement quantities, and the relationship between AEHF and TSAT.

As the AEHF capability and technological requirements expanded and became more demanding, the cost estimates at MS B/C appeared increasingly unrealistic. This lack of cost-estimating realism at MS B/C was based in part on unsubstantiated expectations regarding savings from maximizing the use of COTS technology and other acquisition reform measures. For example, the AEHF bus was intended to use a commercial derivative satellite bus.[4] The program office assumed a very aggressive savings of nearly 50 percent through the use of a commercial derivative bus, and the program office assumed significant savings from reuse of commercial and government-off-the-shelf (GOTS) software.

Yet AEHF, and particularly its payloads, had much more demanding requirements and would be far more complex than the existing Milstar SVs. The technological challenges faced by the contractors included a digital processing requirement for ten times the capability at half the weight compared to Milstar; very high frequency phased array antennas of a type never before used in this capacity; and a major new encryption and secure communications requirement of dramatically increased capability. Nonetheless, at MS B/C, AEHF was optimistically budgeted at only about one-third the cost of Milstar at its MS B equivalent. The underestimated technological

---

[4] A satellite or space bus is the basic space vehicle or structural platform that carries and supports the scientific- or mission-specific payload module. Typical space buses provide electric power, propulsion, communications equipment, attitude control, guidance, navigation and control equipment, and so forth. The payload module is carried on the bus. Since all satellites require these support capabilities, some types of satellites can use the same or similar bus to support different types of customized mission payloads. Use of a common, existing, or commercial bus can reduce non-recurring costs of developing a specialized bus for every type of mission payload. See, for example, Commission on Physical Sciences, Mathematics, and Applications, Space Studies Board, Committee on Earth Studies, National Research Council, Division on Engineering and Physical Sciences, *The Role of Small Satellites in NASA and NOAA Earth Observation Programs*, National Academy Press, Washington, D.C., 2000.

complexity led to substantial unanticipated weight growth in the satellite and payload hardware, as well as a large increase in software lines of code. These specific technical problems were the most important direct causes of cost growth.

Many SMEs were aware of the optimistic official cost estimates during the early phases of the developmental program. More specifically, it was widely believed among many stakeholders and outside observers early in the program that AEHF was significantly underfunded on the order of 50 percent. For example, an OSD Cost Analysis Improvement Group (CAIG) independent cost estimate in 2004 conservatively estimated program costs at 8.7 billion then-year dollars (TY$), yet the program office continued to use the established baseline estimate of 6 billion TY$.

AEHF Summary Findings of Key Program Characteristics

To summarize, the principal common characteristics of AEHF common with many of the other programs experiencing extreme cost growth were

- entering full-scale development (MS B/C) without stable and well-established requirements and program structure, either for the SVs or for the overall system architecture and its relationship to TSAT
- immature technologies and design solutions
- widely recognized unrealistic cost and schedule estimates at MS B/C
- unstable and inappropriate program structure.

C-130 Avionics Modernization Program (AMP)

Summary Overview of C-130 AMP

The C-130 Avionics Modernization Program (AMP) aimed to consolidate and standardize an extensive list of electronics upgrades and modifications for a wide variety of different variants of the existing C-130 tactical transport aircraft fleet. (A cockpit simulator image of the AMP is shown in Figure 2.2.) The program included three major classes of modifications and upgrades: mandated Air Force navigation and safety upgrades; communications, navigation, surveillance, and air traffic management upgrades; and a third category of modifications called Broad Area Review Requirements. The upgrades included installation of fleet-wide radars, aircrew displays, dual autopilots, dual flight management systems, and new communications radios and data links. These modifications were planned to be carried out on 221 C-130s. The C-130 AMP program at its height was one of the largest and most complex aircraft modification/upgrade programs ever planned by the Air Force. Boeing was the prime contractor.

The C-130 AMP emerged from a five-month study of C-130 modernization requirements undertaken by Air Mobility Command beginning in August 1997. The C-130 AMP originally was aimed at replacing and modernizing cockpit electronics and other avionics with modern digital versions on over 500 older Air Force C-130E/H military transport and U.S. Special

Operations Command (USSOCOM) C-130 aircraft, including at least 15 different aircraft model types and subvariants. The original key objective was to consolidate a variety of mandated DoD navigation and safety avionics modernization requirements, the Global Air Traffic Management (GATM) systems, and a wide range of other avionics improvements related to reliability, maintainability, and sustainability requirements. USSOCOM's Common Avionics Architecture for Penetration (CAAP) program for USSOCOM aircraft, aimed at upgrading the same types of avionics capabilities, as well as the unique special operations forces penetration and combat requirements for its highly specialized MC-130E/Hs and AC-130s, were added to the C-130 AMP in late 1999, greatly increasing the complexity of the overall effort.

Figure 2.2. C-130 AMP Cockpit Simulator/Trainer Showing Cockpit Upgrades

SOURCE: Image shared by USAF SOCCOM, via Edwards Air Force Base Media Gallery; no known copyright restrictions.

The C-130 AMP entered MS B in 2001 without a comprehensive understanding of the complexity, technological risks, and challenges posed by the program, and the program also suffered from excessively optimistic MS B cost estimates, in part encouraged by unwarranted claims of cost savings through the planned widespread use of COTS technology as a key element of the acquisition strategy. Program requirements also changed significantly in the early stages of the program. These were the prime causes of extreme cost growth.

C-130 AMP Case History

As noted above, the C-130 AMP, launched in 2001 with a prime contract award to Boeing as system integrator, aimed to achieve a relatively low cost modernization and replacement of the cockpit electronics and other avionics of over 500 aging C-130E/H and other model Lockheed Hercules tactical transports through the use of COTS technology. Like the AEHF, the prospect of extreme cost growth on the C-130 AMP may have been highly probable and predictable even at MS B, given the shortcomings of the information and optimistic estimates at that time.

Some of these factors included early decisions regarding program structure that may have added to the complexity and challenges of the effort. In September 1999, management of the program was moved from the Warner Robins Air Logistics Center (WR-ALC) in Georgia to the Aeronautics Systems Center at Wright Patterson Air Force Base in Dayton, Ohio. Since the many models and variants of these older C-130s had undergone modifications at WR-ALC for decades, this change removed much of the corporate knowledge on the unique modifications. In October, the program was redesignated as an Acquisition Category 1C program (ACAT 1C), putting it solely under Air Force control. By the end of October, the new system program office had removed the preexisting configuration standards, thus permitting greater contractor flexibility in changing the overall architecture, in a move similar to that represented by the TSPR acquisition approach.[5]

These changes facilitated an increased emphasis on use of COTS technology, which was used to justify optimistic cost estimates based on unproven assumptions regarding cost savings. For example, around the time of the July 2001 MS B decision, the baseline assumption was that 100 percent COTS technology would be used for the program's GATM systems, derived from existing commercial airliners, thus providing significant savings. Days after the MS B decision, the prime contract was awarded to Boeing, in part due to its claims that it could incorporate much of the commercial technology from its new B737 airliner cockpit avionics upgrade directly off the shelf into the C-130 AMP. Both WR-ALC and the original equipment manufacturer, Lockheed Martin, expressed skepticism about these claims. Boeing had little experience modifying the C-130, while Lockheed had developed the C-130 and all its major variants. Only two years following the award to Boeing, the estimate of COTS usage had already fallen to only slightly more than 50 percent.

Prior to MS B, USSOCOM's CAAP program was rolled into C-130 AMP, greatly increasing its complexity. This program was incorporated without a complete assessment and full

---

[5] See "C-130 AMP Strategy Shift Will Allow Changes to Aircraft Architecture," *Inside the Air Force*, Vol. 10, No. 42, October 22, 1999; and Amy Butler, "Air Force Opts to Divide C-130 Work Between Depot, Contractor," *Inside the Air Force*, Vol. 10, No. 42, October 22, 1999.

understanding of how challenging the modification and upgrading of the many different versions of Air Force C-130s was, much less the much more nonstandard aircraft from the USSOCOM CAAP program. During this early pre-MS-B period, officials decided to split the production portion of the program between three locations, thus introducing inefficiencies in overhead costs and increasing the difficulty of achieving maximum economies of scale.

In the final analysis, there were two major characteristics and two secondary characteristics apparently associated with extreme cost growth on the C-130 AMP. The two main characteristics were "baked in" to the program structure and assumptions by MS B. First were the large configuration differences among the variants, and indeed among individual aircraft, included in the program, a situation that was not understood and was grossly underestimated. At MS B, the Air Force and Air Force Special Operations Command officially recognized 15 basic configurations among the 519 C-130s slated for modernization. The Boeing bid was accepted on the basis of just six basic configurations that had to be accommodated. One senior DoD official familiar with the program was later quoted in the trade press, after extensive cost growth, saying that once Boeing began the modification effort after MS B, the contractor "found out there were 519 different configurations."[6]

A second main characteristic was known by all program participants from the very beginning of the program at MS B: Because of budgetary and other issues, requirements definition and funding for the C-130 AMP training system were essentially omitted from the MS B baseline estimate. The widespread awareness of this omission is made clear in the program SARs and DAES. And this was not an insignificant omission. The cost of the training system was known at the time to be equivalent to roughly 20 percent of the total program value.[7]

Two secondary characteristics were changing requirements and acquisition strategy. The four main characteristics associated with extreme cost growth are summarized below.

C-130 AMP Summary Findings on Key Program Characteristics

In summary, the main C-130 AMP program characteristics common to many of the other programs experiencing extreme cost growth were as follows:

- underestimation of the complexity of the upgrade effort based on inadequate knowledge of the configuration mix of the targeted fleet of C-130s
- incomplete MS B baseline cost estimate due to the omission of the training system and other factors
- major requirements changes, particularly when the USSOCOM CAAP requirement was added

---

[6] Quoted in Ed McKenna, "A Herculean Upgrade," *Avionics Today*, September 1, 2007.

[7] This situation is clearly indicated in program reporting documents. From the very beginning, the program baseline cost estimate included no estimates for critical training equipment that would clearly have to be procured. Also see trade press accounts, such as Amy Butler, "$1 Billion Increase in C-130 AMP Forces AMC to Overhaul Spending Plan," *Inside the Air Force*, January 21, 2000.

- inappropriate acquisition strategy entailing inadequate oversight of the contractor and unrealistic expectations for cost savings based on COTS.

Evolved Expendable Launch Vehicle

Summary Overview of EELV

The Evolved Expendable Launch Vehicle (EELV, shown in Figure 2.3) program is a family of expendable space launch vehicles derived from existing launch vehicles. It is meant to ensure access to space for both national security payload requirements and commercial satellite requirements. An original key goal of the program was to reduce the cost of space launch by at least 25 percent compared with existing launch systems. EELV encompasses the necessary launch vehicles (rockets), infrastructure, support systems, and payload interfaces to maintain this capability. Currently, the main launch vehicle components of EELV are the Boeing Delta IV and Lockheed Atlas V series of launch vehicles.

Figure 2.3. The Boeing Delta IV Heavy and the Lockheed Atlas V Rocket Launchers

SOURCE: Image shared by Boeing, via Flickr, no known copyright restrictions (left); image shared by Lockheed, via Cape Canaveral Air Force Station, no known copyright restrictions (right).

The U.S. Air Force formulated the EELV program in the early 1990s in an attempt to promote development of a common, affordable family of highly standardized space lifters to replace all existing non-reusable ones. Similar to AEHF, EELV emerged in a budget-constrained environment characterized by the widespread advocacy and adoption of multiple, unproven acquisition reform measures, many of which focused on moving DoD procurement toward greater emulation of commercial development and acquisition processes, commercial-style market-based contracting such as price-based acquisition (PBA), an emphasis on COTS technology, and promotion of CMI of the defense industrial base, all as means of saving increasingly scarce acquisition dollars. By using these and other commercial-like market-based approaches, the goal of the EELV program was to ensure the development of two competing families of space launch vehicles serving both military and commercial customers and aimed at reducing recurring launch costs for DoD by 25 to 50 percent compared with the existing generation of launchers (Titan II, Delta II, Atlas II, and Titan IV).

The most important EELV program characteristics associated with programs experiencing extreme cost growth were excessively optimistic cost estimates and unproven acquisition strategy and program structure. In this case, the acquisition strategy contributed to the unrealistic cost savings thought obtainable through the use of commercial-like contracting approaches and unwarranted optimism regarding the potential of the future commercial space launch market sector.

EELV Case History

From the beginning of the program, the objective was to achieve the mandated minimum of 25 percent savings in recurring launch costs (anticipated savings were officially stated in the 25 to 50 percent range). In the first years, a more traditional acquisition strategy based on competition was undertaken. In August 1995, the Air Force awarded four competing contractors 15-month Low Cost Concept Validation contracts. At the conclusion of these contracts, the Air Force awarded two follow-on competitive pre-MS B concept formulation and risk reduction contracts to Lockheed Martin and Boeing. The original intent was to conduct a more or less traditional down select to one contractor for full-scale development of the space lifters. In September 1997, however, the Air Force began examining the option of continuing competition through development and into production to maintain downward cost pressures on the contractors. The Air Force also began examining alternative acquisition strategies and contracting approaches intended to exploit the anticipated rapid expansion of the commercial launch market against the background of the dot-com boom of the late 1990s.

In October 1998, program officials began implementation of a full commercial-like acquisition strategy and program restructuring. The Air Force decided to retain both contractors to maintain competition, awarding partial developmental contracts under an OTA typically used by the Defense Advanced Research Projects Agency (DARPA)[8] for technology proof testing and

[8] OTA is derived from 10 USC § 2371, which grants authority to enter into transactions other than contracts, grants, or cooperative agreements. The original intent of this provision was to support DARPA projects for the proof testing

prototype development. For the "production" phase, the Air Force awarded two contracts in October 1998 under the newly revised Federal Acquisition Regulation (FAR) Part 12, which significantly expanded the definition of what constituted commercial items for defense procurement purposes.[9] These contracts were commercial-style firm fixed price for the delivery of launch services, not for the procurement of production items.[10]

Thus, instead of directly funding development and procurement of the new launch vehicles in a traditional DoD contracting approach, the Air Force employed alternative contracting approaches (OTA and the newly expanded FAR Part 12) in an attempt to emulate commercial practices and piggyback on the commercial market and industrial base, as promoted by the CMI strategy. These contracting strategies for development and production of the new launch vehicles eliminated most regulatory and statutory reporting requirements, which, along with reduced government program office staffing, undermined the government's ability to maintain oversight of contractor costs and performance.

This approach was adopted because of optimistic expectations about the cost savings achievable through a commercial contracting approach, exploiting anticipated economies of scale, and cost-sharing opportunities with the commercial launch market sector. An important component of the optimistic cost assumptions were projections of a strong commercial market for launch services. In November 1997, OSD approved a strategy of partially supporting two competing contractors to develop launch vehicles and provide launch services based on projections that the commercial and military launch markets would be sufficiently robust to support two contractors. This CMI approach, it was assumed, would provide significant savings for DoD in two ways. First, since projections envisioned a robust commercial market, DoD would not have to pay for the entire costs of developing and operating the launch vehicles. Robust market projections would encourage the prime contractors to finance much of the research, development, test, and evaluation (RDT&E) costs with their own funds, it was believed. DoD could thus behave more like a traditional commercial customer and purchase launch services for military satellites along with the many projected true commercial customers.

---

of new military uses of existing technologies through prototyping. Funds awarded under OTA are generally not subject to the extensive regulations, laws, and other requirements typically imposed on federal contracts.

[9] Since the reforms of FAR Part 12 and the commercial item definitions in FAR Part 2 in the 1990s, which greatly relaxed the definition of commercial items, these regulations have been considerably restricted through a variety of new amendments. The current FAR Part 12 regulations can be accessed online (see References).

[10] This acquisition strategy was mandated by public law and DoD space policy. Congress passed considerable legislation encouraging the commercialization of space and launch vehicles, such as in the Commercial Space and Commercial Space Launch Acts of 1984, 1988, 1998, and 2004. (See Congressional Research Service, *U.S. Space Programs: Civilian, Military, and Commercial*, Washington, D.C., IB92011, June 13, 2006). In 2011, the Air Force began to develop a new acquisition strategy to provide greater insight into contractor launch costs to permit negotiation of more cost-effective contracts for launch services. See GAO, *The Air Force's Evolved Expendable Launch Vehicle Competitive Procurement*, Washington, D.C., GAO-14-377R, March 4, 2014.

Second, DoD assumed that the competition between two providers of launch services would drive down prices, further reducing costs to DoD.[11]

From the very beginning of the program in 1995, when the Air Force sought a minimum of 25 percent savings on launch costs in an environment of increasingly constrained defense budgets, the competing contractors came under unusually heavy pressure to "bid to budget" and submit the lowest plausible bids, without admitting the need to sacrifice any desired capabilities. Optimistic projections of a robust future commercial launch market that could reduce Air Force costs through a CMI strategy became a rationale, whether credible or not, to justify "estimates to budget." In the mid-1990s, even before implementation of a full commercial-like acquisition approach, there were many indications from outside sources that the EELV cost estimates were unrealistic. In 1995, the European and French space agencies reported spending more than $6 billion on the development of the Ariane 5, a similar type of space lifter. Less than two years later, the OSD CAIG reported that the $2 billion cap placed on the EELV development costs during the initial contracts awarded in 1996 was highly optimistic and unrealistic. Further, in November 1997, the Air Force allocated only $500 million to each of the two prime contractors for the competitive development of two new families of launch vehicles using the OTA, assuming the projected rapid growth in the commercial launch market would provide the market incentives for the contractors to finance the full development costs of the new generation of space lifters using their own money.[12]

Given the earlier OSD CAIG cost estimates, some observers expressed skepticism regarding the assumptions of this CMI strategy. In addition to concerns about the optimism regarding the projections of the future commercial launch business base, critics noted that the OTA approach provided no guarantees of specific performance, but only "best effort" by the contractors. However, proponents argued that fixed-price commercial-like contracts based on FAR Part 12 legislation for launch services, combined with competition, would protect the government and result in substantial cost savings.[13]

By 2005 at the latest, it had become clear that the original commercial space launch market projections would not be realized, that there was not enough launch business to support two competing prime contractors, that development costs for the space lifters would greatly exceed

---

[11] See U.S. Government Accountability Office, *Evolved Expendable Launch Vehicle: DOD Needs to Ensure New Acquisition Strategy Is Based on Sufficient Information*, Washington, D.C., GAO-11-641, September 2011.

[12] Expected government launch customers included the National Aeronautics and Space Administration (NASA), the National Reconnaissance Office, and other government users in addition to the Air Force. For an overview of the EELV CMI acquisition strategy and other acquisition reform measures applied to the program, see EELV Program Office, "Evolved Expendable Launch Vehicle (EELV) Program Overview: Affordability Through Innovation," briefing, November 20, 1997.

[13] See Craig Covault, "Delta IV Thrusts Boeing Against Atlas V, Ariane," *Aviation Week and Space Technology*, Vol. 157, No. 22, November 25, 2002; and U.S. Government Accountability Office, *Space Acquisitions: Uncertainties in the Evolved Expendable Launch Vehicle Program Pose Management and Oversight Challenges*, Washington, D.C., GAO-08-1039, September 2008.

original estimates, and that they would not be financed by the contractors, because of the lack of a viable commercial launch market. The fixed price launch service contracts became untenable for the contractors, since the expectation of 75 launches per year that underpinned the pricing had failed to materialize. This led to the need to spread fixed infrastructure costs out over only 15 launches per year rather than the originally projected 75.

At this time, the government Broad Area Review Joint Assessment Team reexamined the entire program and concluded that substantial additional government funding would be necessary to resolve technical issues and complete development of the needed space lifters, as well as cover rising launch costs.

In an attempt to reduce burgeoning costs and compensate for the failure of the commercial market to materialize, in May 2005 the contractors proposed consolidation of their efforts into a single joint entity called United Launch Alliance (ULA) to reduce duplicative overhead and save costs. Other important goals included maintaining assured access to space by supporting two families of space launch vehicles. In December 2006, restructuring into the ULA became official.[14] This restructuring had unclear effects on total Air Force costs, possibly resulting in economies of scale and reduced overhead, but also elimination of potentially beneficial contractor competition.[15]

EELV Summary Findings on Key Program Characteristics

As in the case of both the AEHF and C-130 AMP, the key common program characteristics associated with extreme cost growth were largely predetermined at MS B/C. The most important characteristics for EELV were

- implementation of commercial-like market-based contracting mechanisms and acquisition reform strategies based on questionable market assumptions, which contributed to inadequate government insight into contractor costs, technical issues, and performance[16]
- excessively optimistic cost savings estimates that assumed the ability to leverage the commercial launch market through greater CMI based on unrealistic launch market expectations
- use of cost and schedule estimates at MS B/C known to be unrealistic based on analogous programs and outside independent assessments.

---

[14] The Federal Trade Commission approved the joint alliance in October 2006, subject to compliance with a consent order agreed to by both companies ("FTC Gives Clearance to United Launch Alliance," *Spaceflight Now*, October 3, 2006.

[15] The commercial space launch company SpaceX brought anti-trust litigation against the ULA as early as 2005, arguing that the agreement foreclosed competition. Legal maneuvering was still under way through 2014. See "SpaceX and ULA Go Toe-to-Toe Over EELV Contracts," NASA Spaceflight.com, March 5, 2014.

[16] Again, it is important to note that in the area of space launch systems, both public law and DoD policy encouraged commercialization of launch systems and space activities.

Global Hawk

Summary Overview of Global Hawk

The Northrop Grumman (formerly Teledyne Ryan Aeronautical) RQ-4 Global Hawk (shown in Figure 2.4) is a high-altitude long-endurance (HALE) unmanned aerial jet aircraft system (UAS) whose mission is intelligence, surveillance, and reconnaissance (ISR). Global Hawk can carry a variety of integrated sensor payloads for the collection of various types of high-resolution, high-quality imagery and signals intelligence of targets and other areas of interest for the joint warfighter. There are three key program elements: the air vehicle and sensor payloads, the mission control element or ground control station, and the launch and recovery element. The Global Hawk is the largest, most complex, and most expensive UAS currently in production.

Our assessment indicates there were at least five key Global Hawk program characteristics associated with extreme cost growth on other programs we have examined: (1) lack of clearly defined requirements at MS B/C, compounded by continuing requirements churn throughout development, (2) immature design and technology at formal program launch, particularly regarding payload sensors, (3) optimistic MS B/C cost estimates in spite of available cost histories of prior analogous predecessor systems, (4) inappropriate and unstable program structure, and (5) initiation of production prior to stabilization of the final production design and a reasonable level of design and technology maturity, characterized by the approval of a B/C milestone.

Figure 2.4. RQ-4B Block 40 Global Hawk

SOURCE: Image shared by Northrop Grumman, via Northrop Grumman.com; no known copyright restrictions.

Global Hawk Case History

The initial Global Hawk airframe and systems were developed from 1994 through 2000 as an Advanced Concept Technology Demonstration (ACTD)[17] carried out by DARPA. ACTDs are subject to few of the normal political, regulatory, and acquisition policy reporting and oversight requirements typical of the conventional Air Force acquisition process. Typically, they use the same OTA acquisition approach as was later used on the EELV development contracts. The original goal of ACTDs was to provide a relatively inexpensive means of combining existing technologies and concepts in new and innovative ways, testing out their military utility, and, if found useful, transferring them for use or further development to one of the armed services. Often, prototype systems are developed to meet these objectives, usually designed and developed in a "hobby shop" environment, with relatively little attention paid to issues such as detailed military requirements, producibility, maintainability, and missionization. Thus, transition of an ACTD into a traditional full-scale MDAP is potentially challenging.

The Air Force took over management of the program in October 1998 to conduct the ACTD's demonstration and evaluation phase and carry out a military utility assessment (MUA) using the ACTD prototypes. The ACTD ended formally at the end of 1999, and in 2000 the Air Force authorized a formal development and procurement program, which included almost

---

[17] Later renamed Joint Capability Technology Demonstration (JCTD) programs.

immediate entry into low rate initial production of the basic Block 0 ACTD Global Hawk. This was the first time an ACTD was transitioned into an MDAP. The relatively small program office did not have adequate personnel to manage the challenges of transitioning such a complex program, particularly one with excessive overlap of the development and production phases, an untried acquisition strategy, and changing requirements. At the same time, the prime contractor, Teledyne Ryan Aeronautical, was a relatively small company division that was also unprepared to transition to a MDAP. Teledyne Ryan, although bought out by Northrop Grumman in 1999, took three years to fully staff up for the MDAP.

A fundamental challenge confronting the program was that it entered full-scale development and production after MS B/C with requirements, design features, and technological maturity more appropriate for MS A. Evolving requirements led to a comprehensive redesign of the entire airframe within a year, even though production was already well underway. Adding greatly to the program complexity was the unusually high degree of overlap between the development and production phases, a legacy of the ACTD prototype proof testing approach of multiple repeated and overlapping feedback cycles of develop, produce, and deploy. ACTD proof-test versions of the system were produced and operationally tested and deployed prior to MS B/C. All program phases were undertaken simultaneously with different variants: technology development, engineering and manufacturing development (EMD), production, modification, and sustainment. At the same time, the program was not funded at the level necessary to properly execute such a challenging and complex effort.[18]

A key attribute of Global Hawk associated with the other programs with extreme cost growth was the instability and uncertainty regarding mission and requirements. The original Air Force MUA suggested that the UAS could supplement the Lockheed U-2/TR-1 manned strategic reconnaissance aircraft in the ISR mission area. However, detailed system requirements and operational concepts had not been worked out in great detail prior to the MS B/C decision. Consequently, the original Air Force decision authorized production of an air vehicle largely designed and developed through the DARPA ACTD process with far less Air Force operator input into the requirements than was typical of an Air Force acquisition program. In fact, the

---

[18] Part of the problem with the excessive overlap of the development and production phases was caused by urgent wartime combatant requirements for rapid fielding of the system in support of the Afghanistan conflict. This led to the overseas deployment of developmental prototypes, taking them out of the test program and leading to new performance requirements developed during battlefield experience. While this caused further disruption in the acquisition program, urgent wartime needs obviously trump all other factors. However, this was far from the only or even the most important explanation of the concurrency challenges and instability in requirements. The program was structured from the beginning, before 9/11, to be highly concurrent. More importantly, key mission/role and requirements issues had not been fully resolved when the program passed through the MS B/C process. The simple fact is that the program was not yet ready for full-scale development and production. The urgent wartime need for the platform did not change this fact.

original ACTD developed air frame was incapable of carrying the sensor payload necessary to accomplish the full U-2/TR-1 mission.[19]

The Global Hawk development program was used as a test case for an unusual (for hardware) and ultimately inappropriate acquisition strategy conceived in the 1990s and labeled "spiral development."[20] Unfortunately, little detailed implementation guidance was provided, which made some of the challenges faced by the program more difficult.[21] This acquisition strategy was formulated during the mid-1990s and formalized in May 2003 as the preferred DoD acquisition strategy in DoD Instruction (DoDI) 5000.2.[22] Spiral development envisions an incremental development approach intended to provide great flexibility and responsiveness to user needs during development. Unlike other forms of evolutionary acquisition, a key aspect of the spiral development approach is that the final end requirements for a system are not known at program inception (MS B). With an undetermined final end state, the planning for capability content for each spiral or increment was made even more difficult. Uncertainty over final end-state requirements during full-scale development can lead to requirements instability, and constant shifts in requirements focus can lead to extensive major engineering change proposals during development and the need to retrofit or rework earlier aircraft. Global Hawk development suffered from all these challenges due to the "evolutionary acquisition" approach.

This new strategy posed implementation challenges for the Global Hawk system program office, because no consensus existed among the user communities following MS B regarding the final end capabilities and attributes of the air vehicle and sensors.[23] The fundamental issues of uncertainty were whether the Global Hawk would eventually supplement (and how and to what degree) or completely replace the manned U-2/TR-1, and what capabilities and attributes the Global Hawk had to possess in order to fulfill its ultimate mission, whatever it might be.

This uncertainty over final requirements and ultimate capabilities, combined with extensive program phase concurrencies, all encouraged by the new DoD spiral development acquisition policy, led to significant problems. OSD had not issued any detailed guidance on how to

---

[19] For a detailed account of the early period of the Global Hawk program, see Jeffrey A. Drezner and Robert S. Leonard, *Innovative Development: Global Hawk and DarkStar*—Vol. 2, *Flight Test in the HAE UAV ACTD Program,* Santa Monica, Calif.: RAND Corporation, MR-1475-AF, 2002.

[20] "Spiral development" as defined by DoD in the 1990s was originally developed and commonly used for computer software program development, but never before applied to hardware development.

[21] RAND research indicates that evolutionary acquisition strategies are potentially a beneficial approach to reducing programmatic and technological risks through adoption of a strategy of incremental capability improvement. But our research also suggests that the specific implementation details are of critical importance. One of the most challenging aspects is determining the technological and capabilities content of each increment or program phase. For example, see Mark A. Lorell, Julia F. Lowell, and Obaid Younossi, *Evolutionary Acquisition: Implementation Challenges for Defense Space Programs,* Santa Monica, Calif.: RAND Corporation, MG-431-AF, 2006.

[22] DoDI 5000.2, 2003.

[23] For a more detailed discussion of this new acquisition strategy and its background, as well as the MDAP developmental history, see Lorell, Lowell, and Younossi, 2006.

implement spiral development. Program execution was difficult when few understood the basic construct or how to efficiently and effectively implement it within the regulatory and statutory requirements and restrictions of an MDAP. No clear criteria existed for adding, rejecting, or prioritizing capability enhancements suggested by operational communities and others during development. The engineering staff at the prime contractor was nearly overwhelmed with engineering change proposals. When the Air Force decided to increase the baseline payload capability by 50 percent about a year after MS B, engineers initially underestimated the airframe modifications necessary to achieve this enhanced capability. Initially viewed as a relatively modest airframe modification, attaining this new capability ultimately led to a full redesign of the air vehicle, resulting in the RQ-4B variant.

Further, the extreme concurrency of the program phases also caused major challenges. As in the case of requirements, no clear criteria or assessment methodology existed for allocating limited resources to support operational testing of prototypes, modifying or manufacturing existing assets, and developing new spirals, increments, and blocks. All of these issues led to a complex and often chaotic environment.

Finally, while no directly analogous predecessor system existed to compare the Global Hawk to, we did not find evidence of a rigorous attempt to take advantage of historical cost trends in related system areas. For example, a careful examination of cost trends regarding the TR-1 upgrade of the U-2 developed in the 1980s, and especially its sensors, may have resulted in more realistic cost estimates, particularly in the sensor area. On the other hand, such a historical cost analysis would have been difficult given the failure to clearly designate the ultimate missions, requirements, and desired end capabilities for Global Hawk early in the program.

Global Hawk Summary of Key Program Characteristics

In summary, the principle Global Hawk program attributes associated with other programs experiencing extreme cost growth were

- instability and uncertainty regarding desired end requirements and capabilities
- insufficient program technological maturity at MS B/C
- optimistic cost estimates due to the failure to fully assess historical cost trends in related air vehicle and especially sensor development
- unproven (for hardware) "spiral development" acquisition strategy and program structure; flawed implementation of evolutionary acquisition strategy
- premature approval of LRIP prior to stabilization of the production design, complete overlap of the development and production phases as indicated by a MS B/C approval, further complicated by a new OSD acquisition strategy (spiral development) that advocated concurrency without providing program implementation guidance.

National Polar-Orbiting Operational Environmental Satellite System (NPOESS)

Summary Overview of NPOESS

NPOESS (shown in Figure 2.5) was intended to be an affordable next-generation advanced weather and environmental satellite. Its main goal was to improve weather forecasting in the three-to-seven-day-out time frame for both military and civilian customers. NPOESS was designed to be a polar orbiting satellite using existing technologies and relatively low-risk sensors already under development as prototypes by NASA The overall development program was nominally led by the Air Force, but NPOESS was intended to meet the needs of three different government agencies: NASA, the National Oceanic and Atmospheric Administration (NOAA) within the Department of Commerce, and DoD. The prime contractor and integrator was originally TRW, later bought out by Northrop Grumman.

Figure 2.5. Artist's Rendition of the NPOESS Satellite in Orbit

SOURCE: Image shared by National Oceanic and Atmospheric Administration, via Scientific American; no known copyright restrictions.

NPOESS emerged in the post–Cold War budget-cutting environment. The primary motivation for establishing the joint multiagency NPOESS program was cost savings. Beginning in 1993, in order to reduce duplicative R&D and save costs, Congress and the Office of Management and Budget (OMB) began advocating the merger of DoD's replacement program for the Defense Meteorological Satellite Program (DMSP) and the replacement program for the joint NASA/NOAA Polar Operational Environmental Satellite (POES) program (conducted jointly with the European Space Agency).[24]

The beginnings of NPOESS can be traced to a Presidential Decision Directive to the Department of Commerce, which houses NOAA, and DoD in May 1994, directing them to combine their replacement programs for DMSP and POES into a single joint tri-agency program. Roles and responsibilities were parsed amongst the three agencies, with NASA taking the lead on sensor technology development, NOAA responsible for overall program management and system operation, and the Air Force representing DoD and managing the details of the system acquisition process. In October 1994 a joint integrated program office (IPO) was established along with a joint executive committee representing the three main participants to oversee the program.

As noted above, the overarching objective for this joint approach was to save development, procurement, and operating and support costs through eliminating duplicative RDT&E, personnel, and ground and space assets and overhead costs. The joint approach was anticipated to save an estimated $300 million from FY 1994 through FY 1999, with additional savings thereafter. By 2001, prior to the beginning of full-scale development and procurement, the NPOESS IPO estimated that savings during FY 1994–FY 1999 amounted to $680 million.[25]

Our analysis indicates that four key NPOESS program characteristics are similar to characteristics of the programs we examined that experienced extreme cost growth. The first was the high level of technological risk and uncertainty associated with the planned development and integration of many of the key SV sensors at the beginning of the full-scale development program, combined with the highly optimistic assessment of maturity of the technology at MS B/C. The second was unrealistic and unstable requirements, which was closely related to the first factor and the fourth factor listed below. The third was optimistic MS B cost estimates. The fourth was the use of unproven acquisition reform measures and contracting strategies that reduced government oversight (some of which were similar to those which caused problems on AEHF and EELV). This, combined with a joint management program strategy with no clear single line of authority, greatly complicated program management, leading to a proliferation of performance and technology requirements due to the differing needs and agendas of the primary program participants.

---

[24] See for example Linda D. Koontz, *Polar Orbiting Environmental Satellites: Status, Plans, and Future Data Management Challenges*, testimony before the Subcommittee on Environment, Technology, and Standards, Committee on Science, House of Representatives, Washington, D.C.: U.S. General Accounting Office, GAO-02-684T, July 24, 2002, p. 16.

[25] We do not have access to the historical details of these estimates of cost savings from having a single common satellite development program compared with three separate programs. Therefore. we are unable to assess the realism of these cost-savings estimates. However, such estimates probably contributed to the development of an optimistic official baseline cost estimate at the beginning of full-scale development.

NPOESS Case History

This tension among differing goals and objectives among the three NPOESS joint partners is evident in the period from October 1994, when the IPO was established, and August 2002, the formal MS B/C decision point. For most MDAPs, this period is used for concept and design refinement and technology demonstration and risk reduction. In the case of NPOESS, this period witnessed a significant increase in complexity and technological risk, as the partners' differing objectives led to expanding mission requirements.

For example, in 1996, six major sensors were under consideration for NPOESS. By 2002, this number had expanded to at least 14, of which eight involved new or relatively immature technologies. Because of disagreements over requirements and growing complexity, the formal technical requirements document (TRD) was completed only weeks before the final request for proposal (RFP) was released to contractors. In part because of delays and schedule slips, the contractors received less than one month to complete and return their responses to the RFP, because the government wanted to avoid further schedule slippage.

All these factors contributed to an increase in technological risk and complexity prior to MS B/C rather than a reduction. Much of this was caused by the program structure of joint management and the lack of a single line of program authority and control. Not surprisingly, a study by the National Research Council's Committee on Assessment of Impediments to Interagency Collaboration on Space and Earth Science Missions conducted in 2010 concluded that

> Candidate projects for multiagency collaboration in the development and implementation of Earth-observing or space science missions are often intrinsically complex and, therefore costly, and that a multiagency approach to developing these missions typically results in additional complexity and cost.[26]

Following MS B/C in August 2002, the government awarded a Shared System Performance Responsibility (SSPR)[27] contract for total system integration responsibility. This contract went to Northrop Grumman, which in July 2002 announced an agreement to purchase TRW, the primary integrator up to this point. This was an especially challenging transfer of responsibility from the government to Northrop Grumman, because prior to this point the major sensors had been managed directly by the government under separate more traditional contracts.

This situation was exacerbated by the DoD designation of the program as an ACAT 1C program, which removed it from direct OSD oversight and transferred a very complex tri-agency program to exclusive oversight by the Air Force.

---

[26] National Research Council, *Assessment of Impediments to Interagency Collaboration on Space and Earth Science Missions*, 2010.

[27] The SSPR contracting mechanism was similar to the TSPR contracting concept used on such programs as EELV, and SBIRS, which transferred considerable program responsibility typically exercised by the government over to the prime contractor, and reduced government oversight and control.

Almost from the beginning of full-scale development and production, the program began experiencing serious unanticipated technological and design problems, and the resulting cost growth and schedule slippage. Most of these problems were associated with a handful of the most complex and technologically challenging sensors. The most problematic sensor development program was the Raytheon (Santa Barbara) Visible/IR Imager Radiometer Suite (VIIRS).[28] Other high-risk complex sensors that contributed significantly to cost growth included ITT Industries' Cross-Track Infrared Sounder (CrIS), Boeing's Conical-scanning Microwave Imager/Sounder (CMIS), and the Bell Aerospace Ozone Mapping and Profiler Suite (OMPS).[29] In addition, Northrop Grumman experienced technical challenges with the design and integration of the spacecraft. Even though the sensors' development efforts were de-scoped and other various program restructuring efforts were undertaken, the cost growth and schedule slippage associated mainly with sensor development and integration continued unabated.

Based on the available evidence, the NPOESS MS B/C baseline cost estimates were unjustifiably low because of excessive technological optimism and a failure to heed both the experience of analogous historical systems and independent cost estimates. NPOESS was far more complex and technologically demanding than analogous satellites, such as DMSP and POES. For example, the NPOESS requirement envisioned processing approximately ten times the data as these predecessor SVs. At the time of the NPOESS MS B/C decision, the variants of these predecessor satellites, the POES NOAA-19 and the Geostationary Operational Environmental Satellite (GEO) 15, carried only eight and six major sensors, respectively, while 14 were planned for NPOESS. Additionally, most of the 14 were based on immature technology. The MS B NPOESS estimated average procurement unit cost (APUC) was only about 35 percent higher than the APUC for DMSP in the FY 2000 President's Budget. In addition, the budgets developed for NPOESS at MS B/C were based on a program office estimate that was significantly below other independent cost estimates at the time; the Air Force Cost Analysis Agency (AFCAA) estimate was over 20 percent higher.

---

[28] The VIIRS provides "multi-spectral imagers (which) sample the spectral signatures of features on or near the Earth's surface important for climate science. For over three decades, scientists have depended on this imagery for a wide variety of weather and climate applications." Quoted from National Research Council, Committee on a Strategy to Mitigate the Impact of Sensor Descopes and Demanifests on the NPOESS and GOES-R Spacecraft, Space Studies Board, Division on Engineering and Physical Sciences, *Ensuring the Climate Record from the NPOESS and GOES-R Spacecraft: Elements of a Strategy to Recover Measurement Capabilities Lost In Program Restructuring*, National Academy Press, Washington, D.C., 2008.

[29] The CrIS provides atmospheric temperature and moisture sounding capability. CMIS provides "global microwave imagery and other meteorological and oceanographic wave radiometry and sounding data. . . . Data types include atmospheric temperature and moisture profiles, clouds, sea surface vector winds, and all-weather land/water surfaces." The OMPS provides "measurements of ozone vertical profiles, which are needed to understand and monitor the processes involved in the depletion and anticipated recovery of ozone in the stratosphere." All definitions are *passim* from National Research Council, 2008.

NPOESS Summary of Key Program Characteristics

In summary, the evidence shows that the most important program characteristics of NPOESS that are generally common with the other programs with extreme cost growth were as follows:

- relatively low levels of maturity of technology and high levels of technological risk involved with the design and development of many of the principal sensors selected for NPOESS, particularly the VIIRS, CrIS, CMIS, and OMPS
- unrealistic and unstable requirements
- optimistic cost estimates at MS B
- unproven acquisition strategy and program structure, most notably the SSPR (an offshoot of TSPR) and a joint tri-agency approach, resulting in reduced government oversight without any mechanism for prioritizing requirements and enforcing requirements discipline among three participants with widely differing requirements and objectives.

Space-Based Infrared Systems High

Summary Overview of SBIRS High

The Space-Based Infrared Systems (SBIRS) High program provides four new dedicated military satellites, additional sensors hosted in existing satellites, and associated support systems to fulfill four main early warning, missile defense, and surveillance mission areas. (One of the satellites is shown in Figure 2.6.) It includes sensors placed on existing classified highly elliptical orbit (HEO) satellites, newly developed GEO satellites with multiple sensors, and multiple types of ground control stations and communications links. The prime contractor and integrator is Lockheed Martin Missiles and Space Systems.

SBIRS High is one of the most high-profile and widely studied military space programs in recent years, because it proved to be an unusually troubled program characterized by particularly high cost growth, with overall quantity and inflation-adjusted program cost estimates rising by over 250 percent.

Our analysis indicates that the most significant program characteristics in common with the other programs we examined experiencing extreme cost growth were significant underestimation of design, technology, and integration risk at MS B; unrealistically low MS B cost estimates due to the failure to take into account the costs of analogous systems and the much higher cost estimates of similar contemporaneous proposed systems; unrealistic performance requirements; and the adoption of unproven acquisition strategies that reduced government oversight and control of the program, the most important of which was TSPR.

Figure 2.6. The Number Two SBIRS GEO Satellite Undergoing Ground Testing

SOURCE: Image courtesy of Lockheed Martin; no known copyright restrictions.

SBIRS High Case History

With major development activities beginning in 1996, the SBIRS program[30] was intended to produce the replacement system for the aging Defense Support Program (DSP) satellites, first launched in November 1970 to provide early warning and detection of ballistic missile launches and nuclear explosions. SBIRS fulfills missions for both the Air Force and the national Intelligence Community (IC). Unlike DSP, SBIRS is designed to accomplish four major missions as a key part of a larger system of systems: missile warning, missile defense, battle space characterization, and technical intelligence. The system constellation was originally planned to include four GEO satellites plus a spare, two HEO sensor payloads mounted on

[30] Initially, this program was formally called SBIRS High because an eventual follow-on decision regarding development of a supplemental capability called SBIRS Low was expected. This capability was later moved out of the SBIRS program, and it became generally referred to in the trade press and elsewhere as just SBIRS, even though formal government documents continued to refer to the program as SBIRS High. For simplicity we use the common abbreviated form of SBIRS in the body of the paper.

classified host satellites, and ground support elements. MS B was passed in October 1996. The system was to replace existing DSP infrastructure in the FY 1999 to FY 2003 time frame. The system development and demonstration (SDD, now engineering and manufacturing development or EMD) contract was awarded in November 1996.

SBIRS emerged as an alleged lower-cost alternative after OSD and the Air Force failed to adopt two proposed predecessor programs following the end of the Cold War and the onset of increasingly constrained acquisition budgets in the early 1990s. In late 1990, the Air Force began developing a concept for a sophisticated follow-on system to DSP called the Advanced Warning System (AWS). Over Air Force objections, OSD recommended terminating AWS because of potentially unacceptably high costs and high technical and schedule risk. By April 1991, the Air Force countered by proposing a smaller and cheaper version of its original AWS concept, called the Follow-on Early Warning System (FEWS). OSD argued that this new system concept was also potentially too costly and technically risky. A draft 1991 Defense Science Board task force study finding agreed with OSD's position that an upgraded DSP was the most cost-effective and lower-risk solution, as did an Air Force study of alternatives conducted around the same time.

However, during 1992, FEWS emerged as the preferred Air Force system, and soon evolved into the highest-priority Air Force space program. Nonetheless, DoD finally cancelled FEWS in late 1993. Once again the Air Force responded by proposing another all-new system in February 1994 called Alert Locate and Report Missiles (ALARM) system. However GAO and other outside critics questioned the cost-effectiveness and affordability of ALARM, viewing it essentially as a downsized and stretched out reincarnation of FEWS. Critics argued that both FEWS and ALARM would be far too expensive, with costs estimated at $11 billion or more.

The continuing controversy led to a 1994 DoD Summer Study, which settled on a final architecture for a lower-cost DSP replacement.[31] It determined the ultimate system architecture for SBIRS: four GEO SVs and two HEO payloads.[32] To save money, it placed a heavy emphasis on use of COTS and GOTS technology, recommending use of the BSS 601 standard commercial satellite bus and, perhaps most importantly, reuse of modified versions of the IC's heritage sensors used on all ten of the DSP satellites. In addition, the recommended reuse of existing software from the heritage sensors was intended to reduce costs. Since the IC had presented actual cost numbers for the heritage sensor based on its use in a classified program, the Summer Study participants were reasonably confident about the cost estimates for at least this one critical area. The Summer Study also recommended establishing a joint program office led by the Air Force but including the IC, particularly to facilitate provision of the heritage sensor. Finally, the Summer Study assumed a streamlined acquisition approach similar to that used by the IC would

---

[31] Cited in U.S. General Accounting Office, *National Missile Defense: Risk and Funding Implications for the Space-Based Infrared Low Component*, Washington, D.C., GAO/NSIAD-97-16, February 1997. GAO cites the study as *Office of the Secretary of Defense Space-Based Warning Summer Study*. We were unable to find any final report from this study, but we interviewed individuals who served on the study team.

[32] The HEO sensor payloads would be hosted on a classified satellite.

be adopted to save costs. Many acquisition officials believed a TSPR approach would accomplish such streamlining.

TSPR allowed reduced government oversight of the RDT&E effort and permitted the program office and prime contractor to adopt a more risky design approach and execution strategy ignoring many of the Summer Study's key recommendations. During the final competition to select the prime contractor, the Air Force moved further away from the Summer Study's findings and closer to the earlier preferred all-Air Force ALARM and FEWS concepts. For example, the original ALARM long-term requirement for SV onboard data processing migrated back into the program. In addition, the Summer Study recommendation for using the IC legacy GOTS sensors was ultimately rejected by the Air Force. In late 1996, the Air Force completed its source selection process; Lockheed Martin, the lead contractor on FEWS and ALARM, was the winner. The Lockheed Martin team included a subcontractor for an entirely new sensor based on a paper design rather than the IC heritage sensor. Thus, many of the key assumptions on which the cost estimates generated during the 1994 Summer Study were based had been jettisoned by the time of the final source selection.

The intense competition between two contractors for the final down select during a period of major budgetary constraints pressured both contractors to keep their cost estimates as close as possible to those that had emerged from the 1994 OSD Summer Study, even though those estimates were based on key assumptions regarding a variety of technical issues which were no longer necessarily valid, such as no onboard processing, use of a common modified heritage sensor payload on all SVs, and so forth. Many SMEs involved with the program at that time asserted that the winning contractor's team program cost estimates, as well as the official program office estimates, which were roughly similar to the Summer Study estimates based on completely different assumptions, were extremely optimistic for the existing situation and that the overall program was seriously underfunded.[33]

Once again, as in the case of many of our other case studies such as AEHF, C-130 AMP, and EELV, it appears that the potential for extreme cost growth was predetermined at MS B.[34] At the time of the MS B decision, the official baseline program estimate was optimistic, for two basic reasons. First, the SBIRS MS B estimate, at under 4 billion TY$, was not significantly different than the estimates generated by the 1994 Summer Study. Yet many of the major assumptions and recommendations that underpinned the Summer Study estimates were no longer operative, as noted above. Both the system program office and an AFCAA independent cost estimate at this time placed the program cost in the range of approximately $5.6 billion. Indeed, some critics

---

[33] RAND interviews with program officials and other SMEs, May 2006. Also see Col. Mark S. Borkowski, "Space Based Infrared Systems (SBIRS): Lessons Learned Overview," briefing, updated March 2004.

[34] MS B approval and the commencement of EMD are normally considered to constitute the formal initiation of an acquisition program. See DoDI 5000.02, Enclosure 2, 6c(3), 2008.

viewed SBIRS as not a lower-cost alternative, but merely another reincarnation of FEWS and ALARM.[35]

Second, the actual costs of less complex contemporaneous analogous predecessor systems were higher. The official MS B SBIRS estimate, for more content and capabilities,[36] estimated program acquisition unit costs (PAUC) at $0.93 billion base year 2012 dollars, nearly 14 percent less than the DSP PAUC.

The SBIRS's cost history immediately following MS B seems to confirm the program was seriously underestimated. Shortly after the EMD contract was awarded, its scope was significantly reduced, but the EMD contract value was left unchanged. The final two GEO satellites were removed from EMD and placed into a newly created production program. The absence of an associated reduction in the estimate at completion implies the effort remaining in the EMD contract—designing, building, testing, and delivering the two HEO payloads and first three (no longer five) GEO satellites—was expected by the program office to be more costly than the contract value at MS B.

An additional major SBIRS program attribute associated with the other programs with extreme cost growth is that, contrary to the recommendations of the Summer Study, SBIRS depended heavily on the incorporation of insufficiently mature technologies. Seven years after MS B, GAO reported that several key components and technologies had not reached adequate levels of maturity.[37] Additionally, both the program office and contractors underestimated the complexity and challenge of developing the software and integrating the sensors on the SVs and all system elements together into a single functioning system of systems, particularly in view of the system engineering effort undertaken in support of development.

As the program progressed, the government found itself unable to mitigate the growing technical problems and persistent escalation in costs, even after formal Nunn-McCurdy breaches and recertifications, because of the constraints placed on government action due to TSPR. Initially this prevented the government from directing the contractor to include risk reduction testing. In the early years of the development effort, the government does not appear to have had sufficient insight into the technical details of the program (nor sufficiently trained personnel) to

---

[35] Indeed, at MS B the system program office stated that if the launch vehicles, the replenishment SVs (which were later included in the program), and operations and support costs through 2020 were added to the baseline estimate of less than 4 billion TY$, the total program cost would rise to $10 billion. For a critique of some of the early cost estimates and comparisons, see U.S. General Accounting Office, *Early Warning Satellites: Funding for Follow-on System Is Premature*, Washington, D.C., GAO/NSIAD-92-39, November 1991.

[36] First, there were only five SBIRS High SVs planned, versus ten DSP SVs, thus leading to a situation where fixed costs would be spread out among only half the procurement numbers of SVs. Second, SBIRS included two HEO payloads in addition to the five GEO SVs. Third, SBIRS compared to DSP was tasked to provide much greater capabilities and cover four mission areas rather than just one. For example, SBIRS was expected to duplicate the basic DSP launch warning capability plus add the ability to predict trajectory, the likely impact point, and include a theater missile capability.

[37] U.S. General Accounting Office, *Defense Acquisitions: Despite Restructuring, SBIRS High Program Remains at Risk of Cost and Schedule Overruns*, Washington, D.C., GAO-04-48, October 2003.

competently assess the technical progress of the program and work with the contractor to mitigate problems.

SBIRS Summary of Key Program Characteristics

In summary, the principal SBIRS program characteristics associated with the other programs with extreme cost growth are as follows:

- insufficient maturity of system technologies and design at MS B, and underestimation of integration complexity
- unrealistic performance requirements. While high-level system requirements remained relatively stable, decomposition and flow-down of requirements was not well understood or implemented
- the official baseline estimate at MS B was widely viewed as unrealistic and optimistic; it failed to take into full account the costs of analogous predecessor programs at MS B, and the technological and integration challenges posed by the full SBIRS requirement
- use of TSPR and other acquisition reform strategies that reduced the government's ability to adequately oversee the program, assess progress, and impose corrective actions.

3. Summary Findings and Observations

Summary Overview of the Six Programs

Table 3.1 summarizes the two main categories and five major elements that we identified as the key common characteristics we identified among these six programs experiencing extreme cost growth:

Table 3.1. Two Categories of Common Characteristics of Six MDAPs with Extreme Cost Growth

	AEHF	C-130 AMP	EELV[a]	Global Hawk	NPOESS	SBIRS High
Premature MS B						
Immature technology; integration complexity	√	√		√	√	√
Unclear, unstable, or unrealistic requirements	√	√		√	√	√
Unrealistic cost estimates	√	√	√[b]	√	√	√
Acquisition policy and program structure						
Inappropriate acquisition strategy and program structure	√	√	√	√	√	√
MS B/C (premature MS C)	√		√	√	√	
Unit total (PAUC) cost growth	95%	193%	273%	152%	154%	279%

NOTE: The bottom line of table shows PAUC cost growth for each program, as calculated by RAND, normalized for constant dollars and original planned quantities.
[a] The acquisition strategy became inappropriate only after the anticipated commercial market failed to materialize and the Program Office did not quickly react to the changed circumstances
[b] The initial cost estimates became unrealistic only when the assumed commercial market failed to materialize.

Several interesting points emerge from this table. First, every one of the six programs passed through MS B and entered full-scale development too early. Except for EELV, they all entered MS B without sufficient technology and integration maturity. This problem was often closely linked to requirements instability as well as lack of realism in requirements formulation. Perhaps most importantly, all six programs suffered from unrealistic cost estimates at MS B; in most cases, independent cost estimates and cost estimates of analogous legacy systems suggested that the MS B baseline estimates adopted were much too optimistic. The lack of realism in the cost estimates was also closely linked to the prior two elements, in that either the difficulty and complexity of the required developmental and production efforts were underestimated, or requirements remained unclear or unstable. The lack of realism in cost estimates is a characteristic of all the six programs we examined with extreme cost growth, and therefore must be viewed as critically important.

In fact, five of the six programs ignored the existing documented and well-known costs of analogous predecessor systems. Similarly, five of the six went ahead with formal MS B baseline cost estimates which were known to be unrealistic. This is an indication of institutional pressures within the acquisition system, which tend to encourage optimistic cost estimates early on in programs. Four of the six are space programs. An influential 2003 Defense Science Board study (commonly known as the Young Report) pinpointed the problem of unrealistic cost estimates at MS B as a key cause of cost growth on MDAPs and argued that the space acquisition system is particularly susceptible to this problem. Based on our assessments of all the case studies, this problem, which the Young Report attributes specifically to the space acquisition system, may be more broadly shared by the overall DoD-wide MDAP acquisition system. According to the Young Report findings:

> *Unrealistic estimates lead to unrealistic budgets and unexecutable programs.* The space acquisition system is strongly biased to produce unrealistically low cost estimates throughout the process. During program formulation, advocacy tends to dominate and a strong motivation exists to minimize program cost estimates. Independent cost estimates and government program assessments have proven ineffective in countering this tendency.[1]

Another revealing insight that can be gained by comparing the outcomes in Table 3.1 with the program history summaries is that all the programs used inappropriate, risky, or untried acquisition strategies or program structures. Five of the six programs emphasized acquisition strategies, usually a form of TSPR combined with heavy reliance on COTS, CMI approaches, and commercial-type contracting approaches that reduced government oversight and encouraged optimistic cost estimates. Not coincidentally, TSPR removed what were believed to be unnecessary, counterproductive, and costly constraints imposed on contractors, while simultaneously transferring program responsibility and control to industry in hopes that a more "commercial-like" development approach might prove to be more efficient and effective. Instead, the evidence suggests that this untried and high-risk acquisition approach contributed significantly to the high cost growth, particularly when combined with optimistic assessments of cost savings from the planned heavy reliance on COTS technologies and commercial-like management approaches.[2]

---

[1] U.S. Department of Defense, *Report of the Defense Science Board/Air Force Scientific Advisory Board Joint Task Force on Acquisition of National Security Space Programs*, Washington, D.C.: Office of the Under Secretary of Defense for Acquisition, Technology, and Logistics, May 2003, p. 2.

[2] This does not necessary prove that the TSPR acquisition concept and other acquisition reform measures from the 1990s are fundamentally flawed. Some reform advocates from this era have argued that in most cases TSPR and other acquisition reform concepts were poorly understood and poorly implemented, by both the government and the contractors. The defense divisions of the prime contractors, some have argued, evolved over decades to reflect and operate within the highly regulated and heavily controlled environment of the traditional acquisition system. When confronted with new "commercial-like" acquisition strategies requiring greater responsibility and control exercised

Finally, four of the programs substantially increased programmatic and cost growth risk by implementing concurrent RDT&E and production phases through the use of a combined MS B/C, leading to increased costs during production as assembled systems had to be modified out of station. This strategy is more appropriate when procuring truly off-the-shelf commercial or government-developed items that require little or no further development or modification, or when justified by national emergencies.[3]

Key Lessons Learned and Their Relevance to Today

All six of the programs we examined with extreme cost growth—the "worst of the worst"—began in the mid-1990s, as procurement budgets experienced steep declines following the end of the Cold War, and as the services sought new acquisition approaches to preserve development programs with demanding requirements and high-risk, cutting-edge technologies. The need to drastically cut costs and significantly downsize the acquisition workforce, which traditionally had provided oversight and direction to contractors, led to the adoption of acquisition approaches that were unproven in military acquisition. Many measures aimed at reducing costs by exploiting synergies between the defense and commercial industrial bases, incorporating commercial technologies (especially in the areas of electronics and software), and using commercial developmental approaches, contracting, and best practices. These approaches were often grouped within the broad rubric of promoting civil-military integration, or CMI; commercial market-based approaches to costs and incentives such as price-based (rather than cost-based) acquisition (PBA);[4] and transfer of authority and total "ownership" of the development program, including

---

by the contractor, the long-established structure and culture of the traditional defense divisions of the prime contractors were not equipped to rise to the challenge, or so it is argued by some. Others have argued that, with some types of defense systems and market structures and with appropriately designed commercial-like incentives, contractors would perform more efficiently if given greater latitude. A good example commonly cited is the Joint Direct Attack Munition (JDAM) program. Besides JDAM, other programs have successfully experimented with such approaches, such as the Small Diameter Bomb I program. However, these programs tend to be technologically less complex and have higher production volume, making them more like typical commercial products. At least one acquisition strategy that gained prominence during this period—evolutionary acquisition—merits further consideration, as indicated by RAND research. However, effective implementation of evolutionary acquisition can be challenging. This is particularly true of spiral development, a form of evolutionary acquisition championed in the 1990s that did not require the definition of an end-point capability at program initiation. Flawed implementation of this type of evolutionary acquisition contributed to extreme cost growth on the Global Hawk program.

[3] This strategy was actually widely adopted and considered highly successful during World War II. For example, the Consolidated B-24 heavy bomber entered into high-rate production at the Ford Willow Run facility well before full developmental and operational testing were completed. Finished aircraft rolled off the assembly line and flew directly to a major depot, where they underwent considerable modification before being sent overseas for combat operations. This approach, while possibly costly and wasteful, was considered prudent because it led to a very quick spool up of the B-24 production rate during the wartime emergency situation. For example, see Arthur Herman, *Freedom's Forge: How American Business Produced Victory in World War II*, New York: Random House, 2012.

[4] See Mark A. Lorell, John C. Graser, and Cynthia R. Cook, *Price-Based Acquisition: Issues and Challenges for Defense Department Procurement of Weapon Systems*, Santa Monica, Calif.: RAND Corporation, MG-337-AF, 2005.

design and configuration control (called Total System Program Responsibility or TSPR), over to the prime contractor, in accordance with commercial industrial best practices.

While these commercial-like acquisition strategies contributed to poor performance in these programs, some observers claim that in many respects the jury is still out on their ultimate effectiveness. Some acquisition reform advocates from the 1990s argue that the problem was poor implementation, not the strategies themselves. The emergence of commercial space companies such as Space-X seems to indicate that more commercial approaches may have major benefits. Whatever the truth of the situation, there is no doubt that the strategies were not optimally implemented in programs such as C-130 AMP, SBIRS and EELV. If similar acquisition approaches are tried again, much more care and study must be focused on developing the implementation strategies most appropriate for the defense MDAP environment over future years.[5]

In the early part of the second decade of the 20 century, with more than ten years of war coming to a close, growing domestic challenges, and expanding budget deficits, the United States is clearly entering another period of declining defense budgets and severe acquisition budget constraints. It is critical for DoD to avoid repeating the mistakes of the past by resisting the temptation to turn to unproven acquisition approaches that may claim to deliver complex weapon systems at greatly reduced costs. Instead, DoD should seek acquisition policies that truly address the actual root causes of cost growth on MDAPs.

In May 2009, Congress unanimously passed the Weapon System Acquisition Reform Act (WSARA; Pub. L. 111-23). The law is particularly focused on the pre–MS B period and aims to impose a much higher degree of rigor, discipline, and oversight in the earliest phases of the acquisition process. Based on our analysis, this focus has merit, given the fact that in many of the cases we examined, the key characteristics and conditions associated with programs experiencing extreme cost growth appeared to have been "baked in" to the program by MS B.

However, initial RAND research suggests that the ultimate effectiveness of WSARA remains unclear. It greatly increases the focus on statutory and regulatory requirements for increased discipline prior to MS B, but some initial research indicates that the law's provisions may be too cumbersome and bureaucratic and could add significantly to already lengthy program schedules. Perhaps more importantly, our research suggests the effectiveness of WSARA may be limited because of what it does not focus on: evolutionary or incremental acquisition.[6]

---

[5] For a fuller discussion of these issues, see Lorell, Graser, and Cook, 2005.

[6] As formally laid out in DoDI 5000.02 (2008), the overall preferred strategy is labeled "evolutionary acquisition" and is composed of a series of stepped formal "increments." However, in theoretical discussions of evolutionary acquisition dating back to the 1950s, different terms were often used interchangeably for the same basic strategy, such as *incremental*, *phased*, and even *spiral*. However, later spiral development evolved toward a more narrow meaning applied to software development, and then was incorporated as a specific type of evolutionary acquisition in DoD policy guidance in the 1990s. For an example of an early discussion of this basic strategy, see Robert L. Perry, *European and U.S. Aircraft Development Strategies*, Santa Monica, Calif.: RAND Corporation, P-4748, 1971.

Evolutionary acquisition was vigorously promoted during the 1990s and even as early as the 1950s. This strategy is aimed at greatly reducing technological and programmatic risk at MS B by breaking up programs into a series of discrete stepped or phased improvements and modification efforts or increments, rather than a classical single step to full capability. Unfortunately, in the 1990s DoD advocated a form of evolutionary acquisition called "spiral development," as in the case of the Global Hawk program, which is appropriate for software development but not for major hardware system development programs. We discuss more specific attributes of an evolutionary acquisition strategy below,[7] but first start with the importance of realistic cost estimates at MS B.

Recommendation #1: Ensure That Programs Have Realistic Cost Estimates at MS B

Every one of the six MDAPs we examined with extreme cost growth had formal baseline cost estimates that were either generated without reference to the costs of analogous existing or prior systems or which were known at the time to be unrealistically low. One of the key findings of the Young Report on national security space programs quoted above is that the space system budgeting and acquisition process has strong biases and incentives toward producing unrealistic and optimistic cost estimates at MS B. The same could be said regarding the overall acquisition process for all MDAPs.

We have encountered considerable anecdotal evidence that the process for establishing realistic priorities for the overall Air Force MDAP acquisition portfolio with constrained budgets is flawed. Often, the difficult decisions necessary to prioritize options when budgets are limited are not made. Rather, contractors and government officials may seek to justify unrealistically low cost estimates at MS B to avoid difficult decisions regarding prioritization among important programs and capabilities. Unfortunately, many of the acquisition reform initiatives that were launched in the 1990s era of declining budgets, while possibly useful if properly applied in the appropriate circumstances, were sometimes used to justify launching new programs with unrealistically low cost estimates but without the tools necessary to achieve those savings. In the end, this leads to undesirable consequences, such as ad hoc reductions in capabilities and requirements, reductions in procurement numbers, program stretch-outs that only add to further program unit cost growth, and the need to disrupt other programs to fund these cost overruns. Therefore, it is crucial that the budgeting and acquisition system be reformed in order to

- promote and incentivize more realistic budgetary decisions and planning
- adjust specific program content and approaches to realistically reflect the allocated resources.

Based on our assessment of the case studies, we recommend three key approaches to ensuring the development of more realistic baseline estimates at MS B.

---

[7] For a more comprehensive discussion of evolutionary acquisition strategies, see Lorell, Lowell, and Younossi, 2006.

First, program planners should recognize and incorporate the strong predictive relevance of predecessor programs' costs when establishing new program baseline cost estimates and budgets at MS B. In most of the case studies we examined, more realistic independent cost estimates derived in part from examination of prior analogous systems were undertaken but not used for the program baseline. It is important to recognize the relevance of actual cost data from prior systems.

Second, program planners should maintain a healthy skepticism of claims that new technologies or acquisition approaches can bring about significant reductions in costs compared to prior analogous systems. This is particularly true in the cases where follow-on systems aim for very large increases in overall performance and capabilities compared to the prior systems they are replacing. Government experts must very thoroughly and critically assess exactly how new approaches and technologies could possibly reduce costs while increasing capabilities when such claims are made. And the government must have adequate numbers of the appropriately trained SMEs to conduct these assessments.

And finally, senior acquisition officials should carefully consider alternative cost and risk assessments from independent sources either within OSD or other branches of the government such as the GAO, the CBO, the Congressional Research Service, independent contractors, and others.[8]

Recommendation #2: Embrace Evolutionary Strategies with Comprehensive and Proven Implementation Strategies

Our second major recommendation is that DoD must further refine, embrace, and implement evolutionary acquisition strategies wherever possible to reduce technological risk, lessen the likelihood of turning to immature technologies, and help enforce discipline on capabilities expectations. The fundamental objective of such a strategy must be to reduce and control technology and programmatic risk both prior to MS B and after the beginning of full-scale development. Five of the six programs we examined with extreme cost growth were large single-step-to-full-capability development programs characterized by substantial technological and design immaturity at MS B. At the same time, implementation of effective evolutionary acquisition strategies can be challenging, as was demonstrated in the Global Hawk program. The most difficult problem is determining the appropriate technologies and capabilities for each increment, and rigorously following through. The 1990s concept of evolutionary acquisition

---

[8] The GAO has tracked the problem of poor cost estimates for years. In a 2005 report, GAO observed that realistic baseline cost estimates require better matches between resources and program needs, stable designs, and mature production processes (see U.S. Government Accountability Office, *Defense Acquisitions: Assessments of Selected Major Weapon Programs*, Washington, D.C., GAO-05-301, March 2005). In its most recent report on the subject, GAO discusses DoD progress in this area (see GAO, *Defense Acquisitions: Assessment of Selected Weapon Programs*, Washington, D.C., GAO-15-342SP, March 12, 2015).

using spiral development made this task particularly challenging, because no end-state capability needed to be defined in the spiral development approach to evolutionary acquisition.

Evolutionary acquisition strategies accomplish risk reduction by separating large single-step-to-new-capability programs into multiple increments that are lower-risk and more manageable and, when combined, lead ultimately to the full end capabilities originally sought. While evolutionary acquisition strategies have been widely discussed for years, more comprehensive formal implementation guidance is needed. We make the following recommendations as a contribution to the further development of implementation guidance for a robust evolutionary acquisition strategy:

- Adopt revolutionary technologies only when necessary, such as when required to counter relatively near-term threats. Reemphasize evolutionary acquisition strategies to achieve full objective capabilities through a series of lower-risk program increments or steps. Carefully work out the precise content and capability objectives for each increment prior to launching the development phase of each increment.
- Begin a new ACAT I MDAP only when objective capabilities and goals cannot reasonably be met through a series of smaller, less risky, ACAT II-IV programs.
- Conduct early and comprehensive cost-benefit and risk assessments of requirements and technology, as well as system design and integration, all with an emphasis on affordability.
- Reach consensus on requirements and costs among all stakeholders before determining final formal requirements for the entire program and for each increment.
- Minimize concurrency and overlap within and among specific evolutionary program increments, as well as between major overall program phases.

Comprehensive, detailed implementation guidance for a robust evolutionary acquisition strategy needs to be developed to assist programs in structuring future acquisition efforts. History suggests that evolutionary acquisition strategies can be extremely challenging to implement effectively and efficiently in the real world. Related research suggests that one of the key problem areas is the difficulty in determining the specific content of any given increment or spiral.[9] While much more research and analysis is warranted, some of the fundamental overarching principles that should underpin such guidance are included above. Combined with more realistic cost estimates at MS B, a well-crafted evolutionary acquisition strategy promoting entrance into full-scale development only with mature technologies and designs as well as stable requirements should reduce future instances of extreme cost growth in Air Force MDAPs.

---

[9] For example, see Lorell, Lowell, and Younossi, 2006.

References

"AEHF (Advanced Extremely High Frequency) Series," *Jane's Space Systems and Industry*, June 14, 2012.

Arena, Mark V., Robert S. Leonard, Sheila E. Murray, and Obaid Younossi, *Historical Cost Growth of Completed Weapon System Programs*, Santa Monica, Calif.: RAND Corporation, TR-343-AF, 2006. As of December 9, 2014: http://www.rand.org/pubs/technical_reports/TR343.html

Bolten, Joseph G., Robert S. Leonard, Mark V. Arena, Obaid Younossi, and Jerry M. Sollinger, *Sources of Weapon System Cost Growth: Analysis of 35 Major Defense Acquisition Programs*, Santa Monica, Calif.: RAND Corporation, MG-670-AF, 2008. As of December 9, 2014: http://www.rand.org/pubs/monographs/MG670.html

Borkowski, Mark S. (Col., USAF), "Space Based Infrared Systems (SBIRS): Lessons Learned Overview," briefing, updated March 2004.

Butler, Amy, "Air Force Opts to Divide C-130 Work Between Depot, Contractor," *Inside The Air Force*, Vol. 10, No. 42, October 22, 1999.

———, "$1 Billion Increase in C-130 AMP Forces AMC to Overhaul Spending Plan," *Inside the Air Force*, January 21, 2000.

"C-130 AMP Strategy Shift Will Allow Changes to Aircraft Architecture," *Inside the Air Force*, Vol. 10, No. 42, October 22, 1999.

Commission on Physical Sciences, Mathematics, and Applications, Space Studies Board, Committee on Earth Studies, National Research Council, Division on Engineering and Physical Sciences, *The Role of Small Satellites in NASA and NOAA Earth Observation Programs*, National Academy Press, Washington, D.C., 2000.

Congressional Research Service, *U.S. Space Programs: Civilian, Military, and Commercial*, Washington, D.C., IB92011, June 13, 2006.

Covault, Craig, "Delta IV Thrusts Boeing Against Atlas V, Ariane," *Aviation Week and Space Technology*, Vol. 157, No. 22, November 25, 2002.

Department of Defense Instruction (DoDI) 5000.02, *Operation of the Defense Acquisition System*, December 8, 2008. As of December 9, 2014: http://www.acq.osd.mil/asda/docs/dod_instruction_operation_of_the_defense_acquisition_system.pdf

———, 5000.2, *Operation of the Defense Acquisition System*, May 12, 2003.

DoD—*See* U.S. Department of Defense.

DoDI—*See* Department of Defense Instruction.

Drezner, Jeffrey A., and Robert S. Leonard, *Innovative Development: Global Hawk and DarkStar*—Vol. 2, *Flight Test in the HAE UAV ACTD Program,* Santa Monica, Calif.: RAND Corporation, MR-1475-AF, 2002. As of December 9, 2014: http://www.rand.org/pubs/monograph_reports/MR1475.html

EELV Program Office, "Evolved Expendable Launch Vehicle (EELV) Program Overview: Affordability Through Innovation," briefing, November 20, 1997. As of December 9, 2014: http://www.globalsecurity.org/space/library/report/1997/nov_ovrw.pdf

Federal Acquisition Regulation. As of December 9, 2014: http://www.acquisition.gov/far/

Federal Acquisition Regulation Part 12—Acquisition of Commercial Items. As of December 9, 2014: https://acquisition.gov/far/current/html/FARTOCP12.html

"FTC Gives Clearance to United Launch Alliance," *Spaceflight Now*, October 3, 2006. As of May 20, 2015: http://www.spaceflightnow.com/news/n0610/03ulaftc/

Herman, Arthur, *Freedom's Forge: How American Business Produced Victory in World War II*, New York: Random House, 2012.

Imbens, Guido W., and Donald B. Rubin, *Causal Inference in Statistics, Social, and Biomedical Sciences*, Cambridge University Press, 2015.

Koontz, Linda D., *Polar-Orbiting Environmental Satellites: Status, Plans, and Future Data Management Challenges*, testimony before the Subcommittee on Environment, Technology, and Standards, Committee on Science, House of Representatives, Washington, D.C.: U.S. General Accounting Office, GAO-02-684T, July 24, 2002. As of December 9, 2014: http://www.gao.gov/assets/110/109528.pdf

Leonard, Robert S., and Akilah Wallace, *Air Force Major Defense Acquisition Program Cost Growth Is Driven by Three Space Programs and the F-35A: Fiscal Year 2013 President's Budget Selected Acquisition Reports*, Santa Monica, Calif.: RAND Corporation, RR-477-AF, 2014. As of December 17, 2014: http://www.rand.org/pubs/research_reports/RR477.html

Lorell, Mark A., John C. Graser, and Cynthia R. Cook, *Price-Based Acquisition: Issues and Challenges for Defense Department Procurement of Weapon Systems*, Santa Monica, Calif.:

RAND Corporation, MG-337-AF, 2005. As of December 9, 2014:
http://www.rand.org/pubs/monographs/MG337.html

Lorell, Mark A., Julia F. Lowell, and Obaid Younossi, *Evolutionary Acquisition: Implementation Challenges for Defense Space Programs,* Santa Monica, Calif.: RAND Corporation, MG-431-AF, 2006. As of December 9, 2014:
http://www.rand.org/pubs/monographs/MG431.html

McKenna, Ed, "A Herculean Upgrade," *Avionics Today*, September 1, 2007. As of February 12, 2014:
http://www.aviationtoday.com/av/military/A-Herculean-Upgrade_15357.html

National Research Council, *Assessment of Impediments to Interagency Collaboration on Space and Earth Science Missions*, 2010. As of February 12, 2014:
http://www.nap.edu/catalog.php?record_id=13042#toc

National Research Council, Committee on a Strategy to Mitigate the Impact of Sensor Descopes and Demanifests on the NPOESS and GOES-R Spacecraft, Space Studies Board, Division on Engineering and Physical Sciences, *Ensuring the Climate Record from the NPOESS and GOES-R Spacecraft: Elements of a Strategy to Recover Measurement Capabilities Lost In Program Restructuring*, Washington, D.C.: National Academy Press, 2008.

Perry, Robert L., *European and U.S. Aircraft Development Strategies*, Santa Monica, Calif.: RAND Corporation, P-4748, 1971. As of December 9, 2014:
http://www.rand.org/pubs/papers/P4748.html

"SpaceX and ULA Go Toe-to-Toe Over EELV Contracts," NASA Spaceflight.com, March 5, 2014. As of May 20, 2015:
http://www.nasaspaceflight.com/2014/03/spacex-and-ula-eelv-contracts/

U.S. Code, Title 10, § 2371, "Research Projects: Transactions Other Than Contracts and Grants." As of February 12, 2014:
http://uscode.regstoday.com/10USC_CHAPTER139.aspx#10USC2371.

U.S. Department of Defense, *Selected Acquisition Reports*, Washington, D.C., various systems, various years.

———, *Report of the Defense Science Board/Air Force Scientific Advisory Board Joint Task Force on Acquisition of National Security Space Programs*, Washington, D.C.: Office of the Under Secretary of Defense for Acquisition, Technology, and Logistics, May 2003, p. 2. As of December 9, 2014:
http://www.acq.osd.mil/dsb/reports/ADA429180.pdf

U.S. General Accounting Office, *Early Warning Satellites: Funding for Follow-on System Is Premature*, Washington, D.C., GAO/NSIAD-92-39, November 1991. As of December 9,

2014:
http://archive.gao.gov/t2pbat7/145307.pdf

———, *National Missile Defense: Risk and Funding Implications for the Space-Based Infrared Low Component*, Washington, D.C., GAO/NSIAD-97-16, February 1997.

———, *Defense Acquisitions: Despite Restructuring, SBIRS High Program Remains at Risk of Cost and Schedule Overruns*, Washington, D.C., GAO-04-48, October 2003. As of December 9, 2014:
http://www.gao.gov/assets/250/240492.pdf

U.S. Government Accountability Office, *Defense Acquisitions: Assessments of Selected Major Weapon Programs*, Washington, D.C., GAO-05-301, March 2005. As of May 20, 2015:
http://www.gao.gov/new.items/d05301.pdf

———, *Space Acquisitions: Uncertainties in the Evolved Expendable Launch Vehicle Program Pose Management and Oversight Challenges*, Washington, D.C., GAO-08-1039, September 2008. As of December 9, 2014:
http://www.gao.gov/new.items/d081039.pdf

———, *Evolved Expendable Launch Vehicle: DOD Needs to Ensure New Acquisition Strategy Is Based on Sufficient Information*, Washington, D.C., GAO-11-641, September 2011. As of December 9, 2014:
http://www.gao.gov/new.items/d11641.pdf

———, *Joint Strike Fighter: Restructuring Added Resources and Reduced Risk, but Concurrency Is Still a Major Concern,* GAO-12-25T, 20 March 2012.

———, *The Air Force's Evolved Expendable Launch Vehicle Competitive Procurement*, Washington, D.C., GAO-14-377R, March 4, 2014.

———, *Defense Acquisitions: Assessment of Selected Weapon Programs*, Washington, D.C., GAO-15-342SP, March 12, 2015.